THE DREAM GAME

BOOKS BY ANN FARADAY

The Dream Game

Dream Power

THE DREAM GAME

«««««««««««««««« »»»»»»»»»»»»»»»»

BY

ANN FARADAY, Ph.D.

PERENNIAL LIBRARY
Harper & Row, Publishers
New York, Hagerstown, San Francisco, London

Portions of this work originally appeared in *McCall's.* Grateful acknowledgment is made to Princeton University Press for permission to reprint material from *The Collected Works of C. G. Jung,* edited by G. Adler, M. Fordham, W. McGuire, and H. Read, translated by R. F. C. Hull, Bollingen Series XX: vol.7; *Two Essays on Analytical Psychology,* copyright 1953 and © 1966 by Bollingen Foundation; vol.8, *The Structure and Dynamics of the Psyche,* copyright © 1960 by Bollingen Foundation and © 1969 by Princeton University Press; vol. 10, *Civilization in Transition,* copyright © 1964 by Bollingen Foundation; vol.12, *Psychology and Alchemy,* copyright 1953 and © 1968 by Bollingen Foundation; vol.13, *Alchemical Studies,* copyright © 1967 by Bollingen Foundation,

Designed by Janice Stern

First PERENNIAL LIBRARY edition published 1976

STANDARD BOOK NUMBER: 06–080371–1

76 77 78 79 10 9 8 7 6 5 4 3 2 1

For Peggy and John Benson Brooks

The time that nature has ordained for us to consecrate to repose brings us, with sleep, an accessory more precious than sleep itself; that natural necessity becomes a source of enjoyment and we do not sleep merely to live, but to learn to live well. . . .

Then let us all deliver ourselves to the interpretation of dreams, men and women, young and old, rich and poor. . . . Sleep offers itself to all; it is an oracle always ready to be our infallible and silent counselor; in these mysteries of a new species, each is at the same time priest and initiate.

—Synesius of Cyrene

CONTENTS

Introduction xii

PART I: PREPARING FOR THE GAME

1. DREAMS: THOUGHTS OF THE HEART 3

The Drama of Dreams 7
"Our True Colors" 9
"The Unconscious" 12
Our Day in Depth 15

2. UNLOCKING THE DREAM GATES 19

Our Dream-Thieving Society 22
Saboteurs of Dreaming 24
Reclaiming the Lost Dreamworld 27
Seek Ye First the Guru Within 32

3. KEEPING A DREAM DIARY 37

Instructions for Recalling and Recording Dreams 39
Your Dream Diary 48

PART II: PLAYING THE GAME

4. THE LANGUAGE OF DREAMS—or WHY THE UNIVERSAL LANGUAGE HAS NO DICTIONARY 51

The Logic of the Heart 53
Understanding Dream Symbolism 57

5. EIGHT COMMON DREAM THEMES 67

Dreams of Falling 69
Dreams of Flying 71
Dreams of Nudity 73
Examination Dreams 75
Dreams of Losing Teeth 78
Dreams of Losing Money and Valuables 79
Dreams of Finding Money and Valuables 81
Sex Dreams 84
Identifying Your Dream Themes 90

6. PUNNY THINGS IN DREAMS 92

Dreams Based on Verbal Puns 95
Dreams Based on Reversal Puns 97
Dreams Based on Visual Puns 98
Dreams Based on Puns Involving Proper Names 100
"Dream Cartoons" of Common Slang Expressions and Colloquial Metaphors 103
Dream Puns Based on Body Language 107
Finding Puns in Your Own Dreams 111

7. UNMASKING YOUR DREAM IMAGES 112

Nonsymbolic Dream Characters 118
Symbolic Dream Characters and Images 122
The Three Faces of Dreaming 138

8. ASKING YOUR DREAMS FOR HELP 142

Dream Power as Dream Interpreter 143
Dream Power as *Spiritus Rector* 148
Dream Power as Family Counselor 150
Dream Power as Vocational Counselor 152
Dream Power as Spiritual Counselor 156

Dream Power and Creative Inspiration 160

PART III: GAMES FOR ADVANCED PLAYERS

9. THE HOUNDS OF HELL 169

Topdog in the Manger 172
A Marital Topdog—with Underdog at the End of His Tether 189
The Bulldog Breed 195
Underdogs Turned Topdog 198
Fritz in the Manger 202

10. THE SECRET SABOTEURS 204

Missing the Bus, Train, Boat, or Plane 207
Dreams of Failing to Get Through on the Telephone 213
The Vanishing Stairway, Door, Road, etc. 216
Dreams of Forgetting Lines and Inability to Perform 219
Dreams of Being Unprepared and Arriving Late 226
The Toilet Thief 228
The Hidden Hero 229

11. NIGHTMARES 231

Night Terrors—and Other Demons of the Dark 231
Dreams of Pursuit, Attack, Violence, and Intrusion 236
Animal Nightmares 241
Dreams of Tidal Waves and Drowning 246
Paralysis Nightmares 250
Resolving Nightmares in Waking Fantasy 252
Resolving Nightmares by Requests to Dream Power 255
Resolving Nightmares by Manipulating the Dream 258

12. DEATH AND THE DREAMER 267

Death in the Family 268

Death of a Stranger (and Others with Whom We Are Not Emotionally Involved) 270
Dreams of the Dead 273
The Death and Rebirth of the Self 282

13. THE TRANSFORMING SYMBOL 290

Easy Rider 291
A Clothing for the Soul Divine 295
Trees of Life 300
"The Truth Shall Make You Free" 305

14. THROUGH THE DREAM GATES: ESP AND ALTERED STATES OF CONSCIOUSNESS IN DREAMS 310

ESP and the Dream Diary: the Legacy of J. W. Dunne 313
Underdog's Psychic Radar 316
Shared Dreamscapes 331
Toward Freedom: High, Lucid, and Out-of-the-Body Experiences 336

Appendix 344

Pages from Sara's Dream Diary and Glossary, with an introduction by Sara 344

Bibliography 367

Basic Recommended Reading on Dreams for the General Reader 367
Other Books Cited or Quoted in the Text 368
Further Books of Interest 370

Index 375

ACKNOWLEDGMENTS

I should like to thank Dr. Ross Borden for his advice on my analysis of puns, and John O. Stevens of the Real People Press for allowing me to quote extensively from the works of Frederick Perls.

INTRODUCTION

This book is written in response to requests from all over the world for a comprehensive, step-by-step manual on how to understand and use dreams. My first attempt to bring dream interpretation out of the consulting room into every-day life (*Dream Power,* 1972) inspired many people to start recalling and recording their dreams, but also created a great demand for more detailed instruction on how to interpret them. Many people wrote to me expressing the difficulties they experienced in learning to understand their dreams and asking to join our dream groups; and professionals from many fields—psychiatry, education, religion, and so on—invited us to conduct training workshops in order to get a more tangible idea of how to proceed. As I could not answer all the letters personally, and as we were not able to conduct workshops everywhere, I wrote this book in an attempt to fulfill those needs.

It assumes no prior knowledge of dreams or of *Dream Power,* and starts at the very beginning with help for those who rarely recall dreaming. The first two parts set out the ground rules for keeping a dream diary and understanding the picture language of the dreaming mind, and although much of this will be familiar to professionals, it should offer new insights to those with open minds. Part III, based mainly on research carried out during the past two years, takes the prepared and serious explorer into new depths of experience, and even those who feel they are not yet ready to undertake these stages of the dream journey will find many challenging ideas about the human problems that emerge in dreams.

My approach to dreams evolved gradually from the study groups I initiated in England several years ago after a totally frustrating experience of Freudian analysis which did violence both to me and to my dreams by trying to fit us into a rigid, limited, and dogmatic theory. I had come to analysis from the laboratory, where I sat up night after night for five years monitoring the eye movements of sleeping subjects with electronic equipment, waking them during rapid eye movement (REM) periods and collecting their dreams. It soon emerged that my subjects expected help in the interpretation of their dreams—though this was not the nature of the experiment—and I hoped that analysis would enable me to provide this. I persevered for three years, but finding that Freudian theory was useful in only a small number of cases, I moved on to a shorter Jungian analysis which, although helpful, seemed sometimes to mystify and complicate dreams out of all proportion by looking for obscure universal symbols and "archetypes" when a perfectly simple interpretation made more sense.

It was at this point that a group of us—professionals and laymen who were disenchanted with orthodox forms of analysis—got together to discover for ourselves what our dreams meant. We used suggestions from Freud, Jung, and many others, taking great heart from the work of Dr. Calvin Hall, the American dream psychologist, whose book *The Meaning of Dreams* demystified Freud's ideas and showed how anyone can learn the language of his dreams by looking at them as pictures. By trial and error, we came to the conclusion that the surest guide to the meaning of a dream is the feeling and judgment of the dreamer himself, who deep down inside knows its meaning. So although group members were encouraged to offer suggestions and comments about another person's dream, they learned never to push an interpretation if the dreamer insisted that it did not resonate with his bones. Psychoanalysts and other professionals who claim that their patients "resist" the correct interpretation in order to avoid unpleasant truths are simply turning their pet theories into self-fulfilling prophecies and forcing their own trip on the dreamer. Time and again, when the dreamer failed to resonate with our suggestions, further dreams with recurring symbols proved the correctness of his feeling and the wrongness of our interpretations. The only "correct" interpretation of a dream —that is, an effective interpretation—is one that gives the dreamer a joyful "aha" experience of insight and moves him to change his life in some constructive fashion.

This was the view put forward in the 1920s by the famous American mystic Edgar Cayce, whose information on dreams, ESP, and related subjects came to him from a deep trance state. At a time when educated opinion was accepting the Freudian view, this grade school dropout was saying that dreams can be of many different kinds (including sexual) with many levels of meaning, that lack of interest is the reason for poor dream recall, that only the dreamer knows the meaning of his dream, and that a dream is correctly interpreted when it makes sense to the dreamer, when it checks out with his other dreams, and when it moves him forward in his life. Unfortunately, Cayce's name was unknown to us in England at the time, or I would have included him in *Dream Power* along with Freud, Jung, Hall, and Perls, as one of the great explorers of the dream.

Our dream study groups received a great boost when we discovered the works of Frederick ("Fritz") Perls, the father of Gestalt therapy. I still vividly recall my initial experience of the Gestalt approach when John, my husband, worked on one of his recent dreams at our very first encounter group in London. At the time, I knew nothing of Perls's dream theory, which maintains that all dream images are parts of the dreamer's own personality, and I was puzzled when the group leader, without discussion, asked John to take a seat on one of the two empty chairs facing each other and talk to the characters in his dream as if they were sitting opposite him. He was then asked to move over to the other chair and assume the roles of the various dream images, giving them each a voice, letting them "speak for themselves."

In John's brief dream he arrived home to find our house pulled down. I watched fascinated as John questioned the various dream characters—the house, me, the workmen, and so on—about the demolition, and could scarcely contain my excitement when I suddenly realized that in giving the "demolition foreman" a voice, John was speaking with the accent, gestures, sentiments, and even posture of his own father, who always disclaimed responsibility for his actions by saying it was not his place to question the orders of those in authority. It was a total revelation of the fact that John was allowing our life to be broken up by his first wife, bemoaning the fact, yet doing nothing to stop it, because from childhood he had been conditioned by his working-class father's reiterated conviction that ordinary people are helpless victims of circumstance.

We immediately recognized the power of the dialogue technique and incorporated it into our dream study groups, and John and I still use it in our workshops which we now conduct as a team—but it remains for us only one of many possible ways of working with dreams, not the only way, as some present-day Gestalt therapists like to claim. In fact, Perls's theory that *all* dream images are symbols of forces in the dreamer's own personality is as untenable in the light of modern research as many of the older theories. No single theory can do justice to the richness of dreams, and one very simple fact about them—which almost all psychotherapists neglect completely—is their direct relation to the events and thoughts of the previous day. My years of experimental research made me realize that dreams do not just pick up "residues" from the day in a casual way, but that the whole dreaming process is rooted in the dreamer's present life and concerns. In fact, the dream seems to be a replay of one's day *in depth,* throwing up all the things that waking consciousness was too slow or too preoccupied to catch.

From our dream study groups we evolved our own three-stage method of looking at dreams which helps the dreamer find his own most effective interpretation. A dream can give literal information about things and people in our external world; it can tell us how we *feel* about them; and it can reveal the workings of our own inner world with all its conflicts and healing power sources. Our groups and workshops now are usually very informal, comprising not more than twenty people, and humor is an essential quality for participants, as you will note, from the accounts given in this book. We follow no hard-and-fast rules other than working with one person at a time and taking his dreams through the three levels of interpretation. Anyone can initiate a dream group with family, friends, in schools, and in churches. My purpose in writing this book is to enable you to play the dream game seriously, but with laughter—hence its title.

In fact, a truly scientific approach to understanding dreams is more like learning to play a game than deciphering a code or fathoming the workings of a machine, and many different sets of rules can be followed according to what you want to achieve by playing the dream game. You can play the Freudian game if you want to discover the sexual conflicts of your early years, and it is sometimes useful to know about these; you can play the Jungian game of finding "archetypal" symbols in your dreams that resonate with the world's great myths, and it is sometimes inspiring to do so; you can play the Gestalt game of focusing on topdog/underdog conflicts in the personality—but none of these, nor any other way of working with dreams, is universally true; and we do violence to our dreams if we try to force them into restricting theories.

I agree with Alan Watts that if psychotherapy or personal growth ever becomes divorced from humor, it is a sure sign that something has gone wrong, that we are allowing ourselves to get bogged down in those life-wasting, ulterior-motive "games" which Eric Berne so brilliantly exposed, with devastating humor, in his book *Games People Play.* Fortunately dreams themselves often come up with the antidote to heaviness, by displaying such an outrageous sense of the ridiculous that we are compelled to laugh at ourselves. My hope for *The Dream Game* is that it will enable us all to follow the example of some Zen monks, quoted by Watts, whose most important meditation consisted of ten minutes' guffaw before breakfast every morning at the absurdity of life's destructive games and limiting roles. Played with real enthusiasm, the dream game can be a fascinating and productive work of self-discovery which Jung likened to the sacred *opus* of the alchemist in his search for the life-giving philosophers' stone. But because the name of the game is liberation from bondage, it is also an invitation to the dance.

PART I

PREPARING FOR THE GAME

«1»

DREAMS: THOUGHTS OF THE HEART

That which the dream shows is the shadow of such wisdom as exists in man, even if during his waking state he may know nothing about it. . . . We do not know it because we are fooling away our time with outward and perishing things, and are asleep in regard to that which is real within ourself.

—Paracelsus

What is a dream? Some seem totally trivial, like the idle meanderings of a brain off-duty, and it is not surprising that skeptics say they are best forgotten. But others cannot be so easily dismissed. There are the very frightening ones we wish we *could* pass off as "only dreams," as parents urge their children to do with nightmares, but often their impression is so strong that the memory continues to haunt us for years. There are other dreams of such beauty and joy that we would not have missed them, and still others so vivid that we wonder whether they could be visions of another world or glimpses into some previous incarnation. A few actually predict the future. Is there anything they all have in common?

Although science is still a long way from having any comprehensive understanding of dreams, one finding has emerged pretty firmly from modern research, namely that the majority of dreams seem in some way to reflect things that have preoccupied our minds during the previous day or two. Sometimes this is easy to see, but it is equally true even of those fantastic dreams that seem worlds away from our ordinary life and thoughts, like being chased down the street by a tiger or conversing with a dead person. Dreams express themselves in a special kind of picture language, which I shall be discussing in detail later on; and once this is understood it will be seen that the tiger symbolized someone or something we found frightening the day or so before the dream, while the dead person appeared in order to give tangible expression to some idea he gave us long ago, an idea of immediate relevance to our present life. Dreams reflect not only actual happenings, but also a whole host of thoughts and feelings that passed us by during the day because we were too busy or unwilling to catch them.

In fact, the dreaming mind may be compared to a movie director, picking up things from waking life that need more attention than we have given them, and reflecting on them *in depth* by composing stories in which flashbacks, cartoon-style pictures, and all kinds of other devices are used to express what we are feeling deep down inside about ourselves, other people, and the quality of our lives generally. And this alone, even if we went no further, would be an excellent reason for not merely brushing dreams aside, for is there any human being whose life would not be improved by a little additional reflection? Anyone who seriously believes he has life completely sewn up should consult his friends and relatives for their honest opinion; he will usually find that his gift is simply for making other people carry the burden of his failures and mistakes, and if he continues to do so long enough, they will rebel.

Just what the dreaming process achieves if we sleep right through the dream and never remember it, no one knows. There is probably some basis to the ancient idea of being able to tackle life's difficulties better simply by "sleeping on them," though we can only speculate about how this works. What we do know is that when dreams are remembered, and their reflections-in-depth are understood, a whole new dimension of wisdom and insight is added to life, bringing greater sanity, meaning, and humor into our existence.

One of the commonest questions I am asked is whether the time and effort necessary to understand the picture language of dreams would not be better spent giving straight rational thought to our problems. The answer is that until the dreaming mind brings to our notice all the subtle feelings, vibes, and impressions we have missed during waking life, we are in no position even to evaluate the problems, let alone solve them. A detective with only half the facts of a case at his disposal is unlikely to unravel a mystery, however much reason and commonsense he brings to bear on it. That is why mankind's age-old fascination with dreams was rooted in wisdom, however much it may have been sidetracked into superstition and mystification.

The price our civilization has paid for its heavy concentration on rational thought was nicely brought out in a story told by Jung of a conversation with a chief of the Pueblo Indians called Ochwiay Biano. Jung asked the chief's opinion of the white man and was told that it was not a high one. White people, said Ochwiay Biano, seem always upset, always restlessly looking for something, with the result that their faces are covered with wrinkles. He added that white men must be crazy because they think with their heads, and it is well known that only crazy people do that. Jung asked in surprise how the Indian thought, to which Ochwiay Biano replied that naturally he thought with his heart.

In our culture we not only train our children to think with their heads, but we actively discourage them from listening to their hearts, even though we pay lip service to the power of feelings in human life. When someone behaves out of character, apparently disregarding the dictates of reason and commonsense, we nod wisely and perhaps murmur the famous words of Pascal, "The heart has its reasons which are unknown to the head . . ."—understanding for a moment why it is that so many of our own well-laid plans go astray. But most of the time we have a deep distrust of feeling and emotion, fearing that they will rule our lives if we allow them to flow. In fact, the opposite is the case, for emotion causes trouble only when denied expression. A healthy life is one in which head and heart cooperate without either trying to put the other down. In modern colloquial terms, a person who lives this way is "together."

It is precisely because dreams help us get ourselves together that I have called them "thoughts of the heart," although it would be more accurate to say that dreams show us *some* of our heart thoughts—specifically those we have neglected during the day to our possible disadvantage or danger. A very simple example is a woman who asked me, "I love my husband and my family very dearly, but I have this recurring dream about burying them. What does this mean?" I explained that while no one can seriously try to interpret dreams without knowing a lot about the dreamer and her life situation, the dream suggested that she bore her family a grudge which, if left unrecognized, might flare up into something much more serious. When I questioned her about her life, she replied that she did indeed devote all her time and energy to the family but asserted that surely any normal wife and mother would do the same. Like so many women in our society, she had been brainwashed into accepting a role of service and devotion to the family to the complete exclusion of her own desires and needs, and her dreams showed an urgent warning to get herself together on this, to bring her head (which loved the family) into touch with her heart (which had got to the point of wanting to get rid of them). In the short time available I advised her to try to see herself as a person in her own right, as well as a wife and mother, and to lavish on herself some of the love and attention she had been directing outward to the family. For how can a frustrated and unfulfilled woman make others truly happy?

I have in my casebook countless other examples of dreams revealing this kind of split between head and heart—for example, the successful, sociable businessman who dreams repeatedly of wandering gray, lonely streets in search of a job and friends; the gentle, pacifist student who dreams of himself as an admiral in charge of the fleet; the shy girl who throws herself at men in dreams. In every case the thoughts of the heart are pressing for recognition, and if we ignore them, we shall try to solve our problems in useless and destructive ways. In the three cases I have just mentioned, the dreams are asking the businessman to confront and understand the reasons for his *inner* emptiness, which will simply continue to grow all the time he tries to suppress it by overt sociability; the student is being directed to find some creative ways of expressing a lot of personal drive which his horror of ruthless, competitive society has made him bottle up; and the shy girl is being urged to get in touch with her sexual needs. If such warnings are ignored, not only will the bad dreams continue, but the needs we try to push away will grow and multiply beneath the surface of conscious awareness until eventually they break out in really "uncharacteristic" and destructive behavior.

THE DRAMA OF DREAMS

All these examples illustrate one very important feature of dreams: they express the thoughts of our heart by dramatizing them and often exaggerate in order to make an emotional point. The woman who dreamed of burying her family had not reached the point of actually *killing* them, but she was almost certainly hurting them in small, mean ways like falling ill just at the time of their vacation or accidently ruining a party given by her husband's boss. Conceivably, she could even contract a crippling illness, thereby forcing her family to pay attention to her. The dream expressed her feelings at their most extreme because somewhere deep inside her the forces of health and sanity were urging her to see them clearly and inescapably and *do something about them.* The horror she felt in the dream as she gazed on the dead bodies of her loved ones frightened her sufficiently to ask for help.

In a similar way, the businessman was able to recognize his loneliness when he confronted it honestly, but this was a far cry from his dream picture of wandering the streets in search of a job and friends. At the time we met, however, he was trying to cover his inner emptiness by drinking and socializing, and it is possible that a serious drinking problem could one day cost him his job and his friends. In the meantime, he was getting no real joy out of life. In the group we helped him recognize the split between his head (which assured him he was a fine, successful fellow whom everybody loved) and his heart (which still felt itself to be a lonely, unloved little boy). Once again, the solution lay in learning to love himself in the first instance—the only certain remedy for loneliness.

The gentle student who dreamed of being an admiral was so shocked by his dream that he came to the group seeking an explanation, for he was quite certain he would never want such a thing in real life. "I'd never dream of being one of those pompous buffoons," he said, grinning as he realized his slip. But he did admit to uncharacteristic outbursts of violence that worried him, such as fighting at protest meetings and demonstrations and playing his stereo late at night when his friends were trying to sleep. Of course, his dream was not urging him to become an admiral or chairman of General Motors, but to find an outlet for his vast energy which would be compatible with his hopes for a better world. I do not know what has become of him, but he left the group realizing that there is nothing inherently wrong with energy itself, only in the way it is used.

The shy girl may never get to the point of actually flinging herself at men the way she does in dreams, though her unexpressed sexual needs could drive her into disastrous relationships if not recognized. In dreams we are brought face to face with our heart thoughts at their most passionate. Edgar Cayce knew this and expressed it very vividly by saying that dreams do more than merely provide information and guidance—they produce an "experience" in which the heart pounds, the knees tremble, the spirit sings, and we are *moved* to change our lives.

"OUR TRUE COLORS"

It would be a great mistake, however, to think that dreams are valuable only when they are unpleasant, like the proverbial bitter medicine. Psychologists and psychotherapists have done a disservice both to society and to dreams by their concentration on the idea that achieving greater self-awareness means "facing up" to the fact that we have anxieties, desires, hostilities, and meanness which we are constantly trying to avoid. The point was well illustrated in the famous cartoon strip *Peanuts*, in which two children are playing the game of psychiatrist and patient, with a fee of 7 cents a consultation! Charlie Brown asks his "doctor" girlfriend about the function of dreams, and she replies, "The dreams of the night prepare you for the day that follows . . . it's at night when you're sleeping that your brain is really working . . . trying to sort out everything for you . . . trying to make you see yourself as you really are." Charlie Brown turns away saying, "Even my brain is against me!"

In truth, dreams are even more powerful revealers of hidden talents, buried beauty, and unsuspected creative energy. They urge us to recognize that we are actually a lot nicer than we have hitherto realized. A typical example was the president of a local conservation society who had taken the post ostensibly because he had money tied up in the area and did not want it spoiled. When he joined one of our groups he was astounded to discover from his dreams a deep love and concern for the endangered wildlife around him. His father had been a tough, aggressive businessman who had taught him to despise all forms of gentleness and "sentimentality." But the thoughts of his heart were quite different from the "head trip" his father had forced on him, and once he had got himself together, he was not only able to express his tender impulses without shame, but became much more efficient in the ecology program. "Isn't it strange," one girl remarked after this incident, "how we've all been brainwashed into believing that psychological growth occurs only when we learn something unpleasant about ourselves?"

I am amazed at the inconsistency of many dream psychologists who insist that dreams reveal us in our true colors only when the colors are dark and ugly. When the proverbial Plain Jane dreams of herself as beautiful and successful, her dreams are dismissed as mere wish fulfillments, attempts to compensate for the harsh realities of life. When George, the garage mechanic, dreams of winning the Grand Prix, he is told to come down to earth and stop indulging in Walter Mitty daydreams. But a dream is very different from a daydream, which is largely soporific, and my experience has shown time and time again that such dreams are actually *intimations* of what the dreamer might achieve if he made the effort in waking life.

It is as though the dreams are trying to push across the message, "Look, you are not worthless, inadequate, ugly, unsuccessful . . . with a little effort, you could be much more than you are. Come on, get yourself together . . . move!" Here too the dreaming mind makes its point by dramatic exaggeration, for Jane may never become a beauty queen, nor will George necessarily be a champion. But by enabling us to savor the sweet smell of success, as it were, dreams like these move us to realize our undeveloped potentialities.

A good example was given to us a few years ago by a young man who had dropped out to sail the Indian Ocean. We met him on an island, and after telling us that he never recalled his dreams, he promptly produced one the following night and brought it to us the next day in great excitement. He dreamed that his uncle, long since dead, had left him a painting worth a million dollars in the attic of his old house. After having ascertained that the house itself had been demolished (we always advise taking dreams literally in the first instance!) we asked Ben what his dead uncle had meant to him. "Well," he said, "my uncle always encouraged me to paint, but I never thought I was any good," and he went on to tell us how he had been a printer in East Africa before coming to the island. When we asked if he would still like to paint, he replied that he thought of it constantly but had no confidence in his ability. His parents, he said, had always derided his efforts to do anything creative. We explained that his heart thought otherwise and was urging him to take up the brush. I do not know whether or not he has made a million dollars, but before we left the islands he had started painting and said that he *felt* like a million dollars. As the wise Cayce said, "Dreams are visions that can be crystallized," and in dreams the real hopes and possibilities of the dreamer, not idle wishes alone, are given body and force in order to move us to creative action. After all, the wish is father to the thought, and the thought is the instigator of the deed. Every human achievement (and horror) was first created in the mind and dreams of man.

Dreams certainly do reveal us in our true colors, but these colors can be light as well as dark, gay as well as drab, beautiful as well as ugly. While they undoubtedly do sometimes ask us to face up to problems we are trying to bury, it is much more important, in my experience, to emphasize the positive function of the dream, which is to show us why we get into difficulties in waking life, and why even our best efforts so often fail to bring us the happiness we seek. As Jung put it, "In each of us there is another whom we do not know. He speaks to us in dreams and tells us how differently he sees us from the way we see ourselves. When, therefore, we find ourselves in a difficult situation to which there is no solution, he can sometimes kindle a light that radically alters our attitude —the very attitude that led us into the difficult situation."

"THE UNCONSCIOUS"

Popular psychology sometimes uses the term "subconscious" or "subconscious mind" to describe this "other inside me that speaks in dreams," while professional psychotherapists more commonly refer to "the unconscious" or "unconscious mind." I do not care for such terms, for they carry the suggestion of one or other famous theory of the nature of mind, like the notion that "the unconscious" consists of basic animal instincts, or repressed desires, or the inherited wisdom of the race, or a mysterious realm in which we are all connected to each other like cells in a great Universal Mind. None of these theories is scientifically established, and I know from experience that people often miss important messages from their dreams by approaching them with fixed ideas about the nature of "the unconscious." Some religious-minded psychologists have tried to get around this difficulty by saying that dreams give us valuable messages both from the "subconscious" regions where our minds connect with our bodies and with our storehouse of past memories, and from the "superconscious" regions where we touch universal spiritual forces. Edgar Cayce worked along these lines, and so did the Italian psychologist Roberto Assagioli, founder of the movement known as Psychosynthesis. Such attempts to compartmentalize the mind into "higher" and "lower" seem to me to create more problems than they solve, for often, as I shall show in later chapters, our lives go wrong just because we fail to see the deep spiritual wisdom in things that seem purely animal.

So I have a lot of sympathy with the modern schools of psychology that have suggested dispensing with terms like "subconscious," "unconscious," and "superconscious" altogether. But there are times when it is almost impossible to avoid using *some* shorthand term to describe processes going on in our minds without our being fully aware of them, and for this reason I shall from time to time use the expression "unconscious" in this book, meaning it to be understood in the very broad, simple sense indicated by Jung when he said, "The Unconscious is the unknown *at any given moment*" (italics mine). The thoughts of the heart of which we are unaware in waking life can be of many different kinds, and it is important to pay attention to what dreams have to tell us about all of them.

On the simplest level, dreams can put us in touch with some very mundane things we have missed during the day simply because we were too busy to notice them. If you dream of your teeth falling out, have a dental checkup before going into any deeper level of meaning. Edgar Cayce set the wise example here by stressing that many dream messages arise out of bodily conditions, and a dream of eating fruit and vegetables might simply indicate the body's awareness of a lack of these in the diet.

On a more emotional level, a dream about a disaster occurring in an office where you have just been offered an attractive new job can reveal your heart's concern with vibes you picked up at the interview—the steely look in the boss's eye that contradicted his friendly smile, or a sense of depression pervading the office that belied the glowing promises of a fine career ahead—vibes overlooked by your head in its enthusiasm for money, position, or prestige. Such dreams may or may not warrant overriding the conscious decision already reached by the head, but they do warn us to get ourselves together more than we have by looking into the situation more deeply. It is also possible, of course, for our dreams to pick up good vibes where the head has made a negative value judgment. I once dreamed of a very bossy and overbearing woman scholar as a beautiful young girl, and this helped our subsequent relationship considerably by enabling me to recognize the inner beauty which lay beneath the formidable exterior.

Instances like this occur so often that I sometimes call dreams the "watchdogs of the psyche," constantly on the lookout for vibes, impressions, and facts which the head has missed during waking hours. These watchdogs can also detect the feelings deep within ourselves which we would prefer to push away and forget, like the woman who dreamed of burying her family, or on a more positive note, they can confront us with hidden talents we have been ignoring because our upbringing filled our heads with notions of inadequacy and failure, as happened to Ben and his painting. In every case the dreams may be said to be putting us in touch with "unconscious" processes, because until the dreaming mind confronts us with them, we are for all practical purposes unaware of their existence, although they may have been lying half buried at the back of our heads. And dreams can also put us in touch with processes that are really deeply buried, like physical illnesses which occur as unconscious protests against something unsatisfactory in our lives, although we would never consciously choose to suffer them. Psychotherapists spend most of their time dealing with these deeper kinds of "unconscious" processes and have tended to frame their dream theories accordingly, but for the person who is trying to get himself together by understanding his own dreams, it is important to listen to the watchdogs of the psyche no matter what kind of problem they are bringing to our attention.

OUR DAY IN DEPTH

In all cases, irrespective of what kind of message they have for us, it is important to remember that dreams reflect something in or on our minds *at the time of the dream.* Even when they reveal deep, long-standing problems, or touch on higher transcendental issues, they always show how these things are hitting us *now,* at the present moment in time. And to understand a dream properly it is necessary to see how it relates to some event or preoccupation of the past day or two.

If a dream takes us back to childhood or introduces someone we used to know well but with whom we are no longer involved, it is always because these past events are in some way relevant to a present concern, and it is this reference to the present that constitutes the dream's meaning. While knowing about the past may help a lot in tackling basic life problems, it is essential to know how these problems are affecting our lives at the moment, for it is in the present, in the actual circumstances of life here and now, that we have to get ourselves together. Edgar Cayce, who believed that some dreams give visions of earlier incarnations, still insisted that such dreams must be interpreted in terms of the dreamer's present life situation, for he held that dreams bring us memories of this life or of earlier lives only when such memories are relevant to the needs of the here and now.

The thoughts of the heart have access to a vast storehouse of memories we have accumulated throughout our lives (I have had no personal experience of past lives), and it is uncanny to discover from dreams just how precise some of these memories are. If you take note of the dates of your dreams, for example, you will find that some of them seem to be commemorating an anniversary, by throwing up a picture of an event that happened on that day exactly a year, or sometimes many years ago. This frequently happens with deaths and traumatic stress situations of the past, and it means that the memory is still somehow affecting our lives. It is as if our hearts go on mourning such anniversaries even though our heads have forgotten all about them, and this can sometimes extend to making us physically ill on a certain date each year without our consciously connecting the symptoms with the anniversary. When dreams bring such events to our conscious attention, they are in effect asking us to consider why we are holding on to the unhappiness of the past, and the answer is almost always that we are using the past to avoid having to live fully in the present, perhaps because we fear being hurt again or because it brings some drama to an otherwise empty existence. In my experience such dreams usually show us our fears about the present as well as commemorating the past, and so help us to break free for the future. More commonly, dreams bring us memories of past events to warn us not to repeat the same mistakes in a present situation or to spur us on to more positive thinking about a related present problem. For this purpose they can call up any past event that happens to be relevant at the time, but if a convenient anniversary is at hand they will often make use of it.

The power of the dreaming mind to show us the thoughts of the heart that have passed us by during the day, with all their vast array of associated memories and fantasies, comes first and foremost from the fact that the sleeping brain is not having to pay attention to the outside world. When we are awake we are normally busy doing things, especially in our modern hyperactive culture; but even when we sit down and relax to think about something quietly, the brain is still being bombarded by impressions coming in from the senses, or distracted by thoughts of what we have to do later in the day. In sleep most of these disturbances are removed and the mind is able to give total attention to the thoughts of the heart, and can, moreover, express them dramatically with all the vividness of life itself. Under these circumstances thoughts of the heart have much more chance to claim our attention than they do in waking life.

Over and above this, however, the dreaming mind is able to bypass the prejudices and social pressures that so often prevent us from facing the thoughts of the heart straightforwardly in waking life. While awake, even the most honest and uninhibited of us is conditioned to present the best possible face to the world or try to impress people in certain ways, and while we are doing that we cannot possibly pay much attention to the contrary feelings going on underneath. At a deeper level, we are conditioned right from childhood to view certain feelings as so "bad" that we must never allow ourselves to have them, and our waking minds conform by refusing to admit them even to ourselves. But the dreaming mind cuts right through the pretensions and self-deceptions of the waking mind, riding roughshod over many of our most cherished illusions and showing our feelings for what they really are.

But even this is probably not the whole story about the power of dreams. Brain-wave records indicate that the dreaming brain is even more active than the waking brain, which may mean that it is capable of more work in a given amount of time. My hunch is that the dreaming brain is in a state similar to that produced by the ingestion of psychedelic drugs, when a person experiences things very much more deeply than usual. Computer experts would say that more information per unit time is being "processed" by the brain. Anyone who has taken a psychedelic drug can attest to the fact that subjective time is slowed down so that a minute seems like an hour. This is because more than usual is being made conscious in that minute, which is marvelous on a good trip and devastating on a bad one. Dreaming may show the brain running over the experiences of the previous day or two at a faster rate than in waking life, bringing to our attention all manner of things we have felt or perceived subliminally but have simply not been able to register consciously. This could also account for the dramatic vividness and "exaggeration" of feelings in dreams, both pleasant and unpleasant.

Only further research will show whether this theory of dreaming consciousness is correct, but there can be no doubt that the dreaming mind does offer us a remarkable in-depth review of our lives on a regular night-by-night program, which will enrich and expand us if only we pay attention to it.

«2»

UNLOCKING THE DREAM GATES

In the normal course of things, all of us come awake by way of dream thieving, a psychic discipline of great severity. Each morning, systematically but quite subliminally, we steal from ourselves and sequester every remnant of our prewaking awareness. We have all learned to carry out this exercise in self-impoverishment with a precise and automatic thoroughness. The alarm rings, and instantaneously an axe falls across the continuum of consciousness, sharply dividing awake from asleep . . . an hour after we awake, for most of us the dreams are gone and today has blended into yesterday without interruption or distraction.

—Theodore Roszak

It is amazing how many people still say, "I never dream," for it is now two decades since it was established that everyone has over a thousand dreams a year, however few of these nocturnal productions are remembered on waking. As I verified for myself during my years of experimental dream research, even the most confirmed "nondreamers" will remember dreams if woken up systematically during the rapid eye movement (REM) periods—periods of light sleep during which the eyeballs move rapidly back and forth under the closed lids and the brain becomes highly activated, which happens three or four times every night of normal sleep. Indeed, the brain is working to some extent all through the night, and it is often possible to recall dreams, or vague dreamlike thoughts, even from the periods of deeper sleep in between the REM periods, though at these times both body and brain are much less active, and mental events are correspondingly dimmer.

So it is a very interesting question why some people remember dreams regularly—perhaps several a night on occasion—while others remember hardly any at all under normal conditions. In considering this, it is important to bear in mind that the dream tends to be an elusive phenomenon for all of us. We normally never recall a dream unless we awaken directly from it, and even then it has a tendency to fade quickly into oblivion, sometimes seeming to dissolve in the very process of our trying to grasp it. Arthur Koestler was so impressed by this weird and frustrating phenomenon that he coined a special name for it—"oneirolysis," or dream dissolution. Almost everyone has experienced the sensation of waking up with the knowledge that he has been dreaming, yet recalling nothing but a vague impression of an exciting, depressing, frightening, or joyful experience. A typical example was my six-year-old daughter who recently woke up crying, and when questioned about a possible dream, replied, "I don't remember, but I feel a sadness in my thoughts."

This kind of thing normally happens during the early part of the night, when our sleep is deepest and the REM periods, which occur at intervals of approximately ninety minutes, are quite short, the first lasting only a minute or two, the second only ten minutes or less. If we awaken during the first half of the night, the chances are that we have been engaged in relatively vague non-REM dreaming, or else have just had a very short dream of the more vivid REM variety—in either case giving little dream material for us to grasp as we come into waking consciousness. As the night progresses, sleep gets lighter and REM periods become longer, so that the last REM dream from which we normally awaken spontaneously in the morning can be over an hour in length—but even this tends to dissolve quickly unless we make a deliberate effort to catch it or there is something especially impressive about it that grips the mind.

Given this general elusiveness of dreams, the basic factor that seems to determine whether a person remembers them or not is the same as that which determines all other memory, namely *degree of interest.* Dream researchers have made a broad classification of people into "recallers"—those who remember at least one dream a month—and "nonrecallers," who remember fewer than this. Tests have shown that cool, analytical people with a very rational, unimaginative approach to their feelings tend to recall fewer dreams than those whose attitude to life is open and flexible. Engineers generally recall fewer dreams than artists. It is not surprising then to discover that in modern Western industrial society, women normally recall more dreams than men, since most men are trained to direct their energies out to the affairs of the world, whereas women are traditionally allowed a nonrational, feeling approach to life.

It comes as a surprise to many nonrecallers to be told that their "failure to dream" is due to anything as simple as lack of interest, but dramatic confirmation is often provided by their remembering a dream the very next night after discussing the subject. I get many letters from TV viewers telling me that my appearance on the screen has had this effect. Other people tell me that simply reading an article about dreams was sufficient to break a lifelong habit of failure to recall, and many of our new neighbors—including the mailman—started recalling dreams when it became known that the "dream lady" had moved into the area! The reason for their nonrecall was simply that no one had ever suggested that dreams be taken seriously. On a longer-term basis, however, bias against recalling dreams is not usually disposed of so easily, for a life-style that plays down the inner world of feelings and imagination does not normally change overnight. Moreover, we live in a society where the whole pressure of life and education is in this direction.

OUR DREAM-THIEVING SOCIETY

In modern urban-industrial culture, feelings and dreams tend to be treated as frills and frivolities, which may from time to time be allowed an airing as "women's interests" or in the special area of "mental health," but in general must be firmly subordinated to the hard, practical realities of the factory floor, stock market, office, or household. We pay lip service to the inner life of imagination as it expresses itself in the arts, but in practice relegate music, poetry, drama, and painting to the level of spare-time extras, hobbies and "recreations" that are valued mainly by the extent to which they refresh us for a return to work. We discourage our children from paying much attention to anything that might detract from the serious business of studying for exams or making a living in the "real" world of industry and commerce. There is evidence that in other cultures less frenziedly extroverted than our own, both in earlier ages of history and in nonindustrialized parts of the world today, dreams have been much more commonly and generally recalled than they are in our urbanized society.

A great many people are beginning to recognize that our present world's emphasis on the practical and material has gone too far, to the point where all our efforts are simply not helping us to achieve their supposed objectives of happiness and fullness of life. Many young people, in particular, are protesting that our so-called down-to-earth realism is not truly either realistic or practical; but is something like an obsession, which all too often produces breakdown, alcoholism, and violence, and could indeed lead to global disaster through pollution and the rape of nature. There is a great upsurge of interest in mysticism, meditation, and the inner life, including dreams; and Theodore Roszak is a spokesman for this movement in his book *Where the Wasteland Ends,* in which he accuses modern urban-industrial culture of "dream thieving":

> We in the contemporary west may wake each morning to cast out our sleep and dream experience like so much rubbish. But that is an almost freakish act of alienation. Only western society—and especially in the modern era—has been quite so prodigal in dealing with what is, even by the fictitious measure of our mechanical clocks, a major portion of our lives. And what becomes of those who break under the strain of such a spendthrift sanity—for many do? Does not our psychiatry return them to their dreamlife to recover a portion of what they have squandered? Yet, ironically, "normal" men and women rise every morning of their adult lives to dose themselves on caffeinated drinks whose purpose is to expunge their dream experience. Many are driven by the demands of their work to intensify the waking state and prolong it well into the night by a punishing regimen of pep pills and "uppers." The business life and politics of the artificial environment are transacted exclusively by men and women in just such a condition of exaggerated alertness. Sad victims of a grueling addiction—yet they enjoy high regard for being practical, productive, "wide awake."

Roszak argues that we should not use dreams just on special occasions, as psychiatrists do, but should change the whole temper and direction of life so as to allow more awareness of the "dark mind," the different state of consciousness that takes over during sleep, in which we get an entirely new perspective on the world and touch the roots of creative imagination. He reminds us that prescientific cultures and most high civilizations of the past paid special attention to dreams, with the result that they enjoyed a depth and richness in life unknown to us in the modern West. In *Dream Power* I suggested ways in which a new sanity could be brought into our society by adapting our major institutions to take account of the dreaming process, not only families but also churches, schools, colleges, and even work organizations. To those so-called realists who ask how we can afford the time for it, I would answer that in the present plight of our society we cannot afford *not* to do so.

SABOTEURS OF DREAMING

While it is true that our urban-industrial society is responsible for dream thieving on a vast social scale, I have detected in the course of my work with dreams a subtler saboteur who steals the dreams even of those who profess a sincere love and interest in the psychic depths of man. For example, many psychoanalysts and psychotherapists sabotage their patients' dream recall by showing a special interest in only one particular kind of dream to the exclusion of all the others. My own Freudian analyst made it quite clear to me that he valued my sex dreams above all others, and there is evidence that Jungian analysts often fall into the trap of showing undue interest in dreams of a mythical or archetypal nature. Patients detect these biases very quickly and soon learn to stop mentioning, or even recalling, their "uninteresting" dreams, with the result that their dream life becomes sadly diminished. The analyst then is able to announce triumphantly that his particular theory of dreams is corroborated.

I find the same phenomenon occurring among members of some religious and occult groups who are interested in dreams as supernatural prophecies, reincarnational visions, or extrasensory perceptions. I was puzzled at first by the frequency with which people in such groups complained to me that they had had no dreams for months. On closer questioning, however, I found that what they really meant was that they had had no dreams of the kind they were interested in for months. They had been dismissing all "ordinary" dreams as irrelevant meanderings of the sleeping brain, unworthy of serious consideration. They simply do not think of them as dreams at all and are mystified when other people say they dream frequently. We have also met people of various sects who dismiss all dreams except those in which Jesus Christ, Krishna, Mohammed, or some other great teacher appears, so it is not surprising to find that such people believe they dream only once or twice every blue moon. Persistent dismissal of ordinary dreams in this way actually reduces the amount of dreaming the person *remembers,* leading to the apparent paradox that people who profess a strong interest in dreams often class themselves as nonrecallers. It is almost as if there were some kind of automatic selector mechanism in the brain that prevents people with strong preoccupations like this from remembering the dreams they consider "unnecessary."

Because I am convinced that we profit by being open to *all* messages of the dreaming mind, whether they are concerned with our diet, our relationships, or spiritual disciplines, I am sorry to discover that some mystical teachers who have captured the imagination of many young people today are inclined to dismiss ordinary dreams as unnecessary distractions from the pursuit of enlightenment. Some Eastern gurus are culprits here, but the case that has attracted widest attention is that of Don Juan, the old Mexican Indian sorcerer whose teachings have caught the imagination of Western society through the writings of the young University of California anthropologist Carlos Castaneda, who became his disciple in the early 1960s. In his third book, *Journey to Ixtlan,* Castaneda reveals that Don Juan is indeed very interested in dreaming, but for one purpose only—as an aid to the development of mental power. An important step in this, he says, is to learn to become conscious in a dream that you are dreaming and then learn to manipulate the dreamworld.

According to Castaneda, Don Juan calls only this *dreaming,* as contrasted with ordinary dreams, which he dismisses as unimportant. He recommends that Castaneda cultivate the art of *dreaming* by learning to look at his hands every time he catches sight of them in an ordinary dream, thereby (hopefully) inducing the conscious realization of being in a dreams when the dreamer has learned the art of stabilizing this consciousness, Don Juan says, he has control over his dreamworld and "can act deliberately, he can choose and reject, he can select from a variety of items those which lead to power, and then he can manipulate them and use them, while in an ordinary dream he cannot act deliberately." He continues, "In *dreaming* you have power; you can change things; you may find out countless concealed facts; you can control whatever you want."

Now this special kind of dream consciousness, which dream experts call *lucid* dreaming, is indeed an extremely interesting phenomenon which, I believe, can be the means of attaining higher states of consciousness, but my own researches show that it cannot be forced to appear before its time and that the lucid state begins to occur naturally *after* the dreamer learns to use his ordinary dreams to clear up his life problems.

The essential point, however, is that Don Juan is not really concerned with the problems of everyday life at all; his interest lies in dimensions of mind far beyond those we in civilized society can even grasp. But for those of us, including Castaneda, who have not left the world and its problems behind, and may never do so, then our immediate help may lie in other directions. On this we must use our commonsense. Eastern gurus who often urge us to dismiss all dreams as passing fancies are extrapolating from their own special position of spiritual development, apparently overlooking the fact that most of us are very far from that enlightened state. As I shall show later, dreams are invaluable in mapping out our spiritual progress, whatever our special discipline may be, and in pointing out internal blocks that sabotage our efforts. For all of us living in the ordinary world, the dreaming mind is capable of adding a whole dimension of wisdom to life which is equivalent to the discovery of an inner guru within each one of us, and it would be a great mistake for us to be dissuaded from using this source of insight and wisdom by taking the remarks of Don Juan or any other master out of context.

RECLAIMING THE LOST DREAMWORLD

The most effective of all stimulants to dream recall is the practical discovery of how valuable dreams can be in life, and we have come across many cases of chronic nonrecallers turning into vivid recallers after attending a dream workshop. In a recent college course several students started recalling their dreams when other members of the group dreamed about them in embarrassingly truthful ways. This gave rise to reciprocal dreams, which in turn led to very effective group encounters. Some students left college after working on their dreams, realizing they were there only at their parents' insistence, while others who had considered dropping out to live with the Indians came to see how much they enjoyed college life.

Group work also has the effect of stimulating greater recall even in those who have been in the habit of recalling dreams fairly regularly. Walter, who attended a workshop we conducted in Pennsylvania, wrote later to say, "There was a period of about a month or so, shortly after returning from the dream weekend, that I was getting up two to three times a night to record dreams. My apartment mate started calling me the All American Dream Machine." I constantly urge the formation of small dream groups among friends and colleagues, as I know from experience that this is by far the most effective way to keep the dream life active and dynamic.

Any nonrecaller who has followed me so far in this book will probably have been stimulated to remember at least a few dreams—which is sufficient to start work along the lines described in Part II. If your dreams continue to elude you in spite of everything I have said, then you are probably being bugged by some unperceived bias at the back of your mind which needs to be brought out into the full light of consciousness, and for this purpose I would suggest the following experiment:

Place two chairs opposite each other about five feet apart and sit in one of them. In imagination, place "your dreams" in the chair opposite and talk to them. Say, in your own words, something like this:

"Dreams, why don't you come to me?"

Then move to the other chair and take the part of the dreams, giving them a voice, letting them speak for themselves through your vocal cords. Try not to feel embarrassed about talking to yourself. Many people are indeed very self-conscious the first time they try this, but you will be surprised at how quickly the embarrassment passes. Let the answer come naturally, saying the very first thing that comes into your head. Try this experiment before reading further, as I am now going to give some of the answers other people have received from their absent dreams, and I do not want them to influence your own answer.

These are some typical answers given by chronic nonrecallers when they spoke in the role of "dreams":

"We don't come to you because you are a very busy person and we don't want to disturb your sleep."

"You don't consider us important, so why should we bother to come to you?"

"You have life nicely sewn up at the moment, and we don't want to spoil that for you."

"You have enough problems: we don't want to bring more."

"We are so sad and unhappy we would make you cry all night."

"You are confused enough already: we would only make you worse."

"We are not as exciting as we used to be, and this might depress you."

"You are too busy to be bothered with trivialities like us."

"You don't understand us when we do come, so what's the point?"

"We don't want to frighten you."

If you are a chronic nonrecaller and your answers are anything like these, then you are probably the victim of what Fritz Perls called "catastrophic expectations," for all these fears are based on misapprehension. In the first place, as I have already indicated, the process of dreaming is in no sense inimical to getting a good night's rest or sleep: on the contrary, it is an essential part of it. Nor is there the slightest reason to believe that jotting down a dream during the night or in the morning will interfere with the day's work. It is just as much a misapprehension that dreams can make trouble for us or bring unhappiness, for the simple reason that they reveal only what is already there. If you fear your dreams will disrupt your life, you are merely articulating a half-conscious anxiety that your life is in danger of being broken up by anything at any time, a precarious and unsatisfying mode of existence. Better by far to listen to the dreams before it is too late.

The pictures thrown up by dreams may conflict with the self-image we have cultivated since childhood, but they do not lie, and if we try to ignore inconvenient truths about ourselves or others, we do so at our peril. The fear that dreams will bring us unwelcome or frightening fantasies is very common and was illustrated vividly by a woman viewer who called in with a dream when I appeared on a TV breakfast show in San Francisco. "I am in my late seventies," an embarrassed voice began, "and my neighbor is in her late sixties. We have been friends for years and have always greeted each other with a loving kiss. However, she has now begun to kiss me like a man, and I want you to tell me, doctor, how I can stop myself dreaming, because I'm sure my dreams will take the situation further!" I replied that the dreams she feared had already taken place in waking fantasy and that it would be much better to allow them to flow openly, rather than spend time and energy trying to keep them down. I told her that we all have sexual fantasies, homosexual and otherwise, and we should look at them with humor rather than disapproval. I hoped my own sexual energies would flow as strongly when I reached her age!

The idea that having or even enjoying sexual fantasies will plunge us into debauched behavior is another catastrophic expectation which needs only to be viewed in the clear light of consciousness to collapse. The truth is that we are much more likely to be led into sexual difficulties and awkward situations if we resolutely thrust the feelings and fantasies aside. In the same way, we are often fearful of facing the emotions of anger, resentment, greed, and self-pity, because of the unexamined notion at the back of our minds that merely to admit the feeling is to be forced to act on it, when the reverse is the case. If dreams show you killing your spouse or your boss, there is no rule that says you have immediately to file for divorce or change your job, any more than a sexual fantasy about a neighbor has to be translated into action. You can still take rational thought about what is convenient, expedient, or what you consider to be morally right, and wherever possible you can talk over your newly recognized feelings with the other people involved. If you fail to recognize what your feelings are, however, you are likely to cause real damage to yourself and others by disrupting their lives or by becoming unaccountably ill or depressed yourself.

Yet another catastrophic expectation is that the effort to understand dreams and utilize their messages will prove so time-consuming that we shall be forced to neglect important duties. My own experience, and the experience of everyone with whom I have worked, is that attention to dreams usually reveals enormous amounts of energy being wasted through insufficient knowledge of ourselves and others—so much so, that the effect of spending time on learning the language of dreams and interpreting their messages is to produce a day *less* pressured than before, rather than more so.

The result of facing up to your catastrophic expectations and seeing through them should be to unlock the dream gates the very next night, but it may be useful reinforcement to round off the "dialogue with dreams" with something like a formal apology for misunderstanding their motives. Put them in the chair again and say something like this:

> "Dreams, I didn't realize how important you were. If you will come to me, I promise to write you down and try to understand you."

Or, if you are afraid of nightmares, try this approach:

> "Dreams, could we make a deal? I can't cope with you if you frighten me too much. If you would send your message in a friendly way, I promise to listen."

If the dialogue technique fails to work at once, then it may be that you have not really convinced yourself. *Do not give up.* Try repeating the experiment and just before you fall asleep issue a firm invitation to dreams to come to you. Make plans to record your dreams properly along the lines described in the next chapter. People frequently complain that dreams are not coming to them as much as they would like, and when I ask if they are writing them down, they plead that other obligations have been too pressing recently—to which I answer that your dreaming mind knows very well how seriously you are really taking it and reacts accordingly.

SEEK YE FIRST THE GURU WITHIN

I am sometimes asked whether it could be dangerous to stimulate the recall of dreams. Once we have opened the dream gates, might not our dreams burst forth and overwhelm us? Or perhaps the lure of the dark mind is so great that we might be sucked into morbid introspection and neglect our duties in the external world. My answer is that dreaming is as dangerous as living—no more, no less—but it has also been my experience from working with thousands of people all over the world (including those who left psychotherapy because they were getting no help from it and others who have actually been sent to dream workshops by psychiatrists) that such fears are just another example of catastrophic expectations, with no foundation in fact. As the main culprits in spreading such fears are psychiatrists and psychoanalysts, I sometimes wonder if their real concern is with possible loss of business or professional prestige.

I am appalled at the way we allow experts to rule our society, when a little homework in the various fields would render their services quite unnecessary most of the time. In his latest book, *Tools for Conviviality,* Ivan Illich attacks what he calls the "imperial professions" which encourage dependence and claim monopolies over particular areas of our lives. He sees modern medicine, including psychiatry, as a major threat to health, concerning itself principally with the preservation of the sick life of medically dependent people in an unhealthy environment. He adds that the real danger lies not in too little medical care, but too much.

Fortunately, there are signs that the tide of public opinion is turning on this subject. In the health field in particular the whole question of the expert is under scrutiny—for example, women are being encouraged to liberate themselves from their gynecologists and reclaim their bodies as their own, and in time the dying may be able to take their deaths into their own hands by deciding for themselves the time and place to die. The day of the guru, if not over, is coming to a close as more and more people take responsibility for their actions and their lives, calling on inner resources rather than rushing off to an expert every time the going gets a little tough.

I would never recommend anyone to go to a psychiatrist or psychotherapist for individual treatment except in the most extreme emergency, for what normally ensues is a lengthy course of therapy that eventually becomes a cop-out from living. The late Dr. Eric Berne, founder of Transactional Analysis, put the matter bluntly by saying that most patients in psychotherapy spend years "making progress" (which means learning to live a little better with their neuroses) instead of changing their lives and getting well. When any therapist becomes part of your daily life and you rely on him for advice and support, then you have abandoned responsibility for your life and are in danger of becoming a walking automaton. For this reason, many professionals (like Berne, and Carl Rogers, Perls, and others) have themselves turned to group work for the majority of their therapy. This has the additional advantage of providing a built-in safeguard against the therapist who impresses his own life values on the patient, something that is almost inescapable in individual treatment, however much the therapist claims to be exercising a nonjudgmental function. My own recommendation to people who feel in need of help is to join a humanistically oriented therapy group.*

In the case of dreams, I have never known anyone to be overwhelmed or sucked into morbid introspection, although it is possible sometimes to become depressed or frightened by them. I do not see anything alarming in this, for we are often made anxious and upset by events and traumas in our everyday waking lives without falling into a decline or nervous breakdown. How many of us are really so weak and vulnerable that we cannot stand a little disturbance from time to time? In the majority of cases it is a mistake to treat anxiety and depression as "symptoms" to be removed as quickly as possible; they occur for a reason, usually to slow us down or cause us to rethink our lives, and the answer is not readjustment to the same life-style that caused them. If we listen patiently to our dreams and the messages they contain—even if they do sometimes disturb us—they will eventually lead us to health by throwing more and more light on our problems and pointing the way to constructive solutions. How much better to take advice from the other half of yourself than from another person.

*The Association for Humanistic Psychology (325 Ninth Street, San Francisco, California 94103) regularly publishes a list of centers where such groups are held.

I am particularly disappointed when members of our dream workshops say they did not work on a worrying dream during the week because they knew they were coming to the group or to their therapist, for the whole purpose of the exercise is to cultivate confidence and reliance on one's own powers. Even nightmares have a positive function, as I shall be showing in a later chapter. I want to see people claim their souls for their own, and I hope this book will help everyone who reads it to find and trust the only expert of any real importance—the guru within, who will not send us more in our dreams than we can handle at any given time. For example, a friend of mine had been working intensely with dreams over a long period and was beginning to neglect her affairs in the external world. Far from being sucked down into the morasses of the dark mind, her dream recall stopped abruptly. She was distinctly annoyed and turned to the dialogue technique to find the reason, whereupon she received this reply from her "dreams": "You must get a job. Your money is almost finished and you can't live on air, so we are going to stay away until you establish yourself firmly in the world again."

Dreams can also stop coming if you are locked into some way of interpreting them that has ceased to be useful to you. This happened to Charlie, a teacher in Georgia, who was studying psychology and had worked with his dreams for several years along Jungian lines, interpreting them in terms of archetypes. In a college paper, Charlie wrote that "eventually this approach became stale. There was a discouraging sameness in the archetypes from day to day—there was no reconciliation—and the sameness also pervaded my interpretations. Eventually I stopped dreaming so much, neglected to remember my dreams, and became despondent about the whole business. I knew I was not doing them justice. This went on for some time, and I professed to have become bored with dreams. . . ." The paper goes on to describe how Charlie's interest in dreams, and his recall of them, grew to life again when he discovered Fritz Perls's Gestalt technique. The moral of the story is not that the Gestalt approach is necessarily correct, or better than the Jungian way, but merely that Charlie needed something new in his dream life. Jung himself wrote that "it is important not to have any preconceived, doctrinaire opinions about the statements made by dreams. As soon as a certain 'monotony of interpretation' strikes us, we know that our approach has become doctrinaire and hence sterile."

"The alternative to recalling and interpreting dreams," said Edgar Cayce (as reported by Harmon Bro in his book *Edgar Cayce on Dreams*), "is not always pleasant. Individuals cannot expect to drift forever. If they do not puzzle out their identity, and the direction of their lives by the aid of their dreams (which he said every normal person should try to do), then they may be brought, by the relentless action of their own pent-up souls, into some crisis which requires that they come to terms with themselves. It may be a medical crisis. It may be the end of a marriage or of a job. It may be depression or withdrawal. . . ." In other words, the real danger lies not in encouraging dreams to come to you, but in blocking them out of your life.

«3»

KEEPING A DREAM DIARY

I never travel without my diary. One should always have something sensational to read in the train.

—Oscar Wilde

The first essential in playing the dream game, for good and poor recallers alike, is to make a regular habit of getting any dreams you remember down on paper. Even dreams that seem boring may well turn out to have *something* to say, if only by making you familiar with the way your mind works while you are asleep, and so casting light on the understanding of other dreams. Nothing is too uninteresting for use, particularly in the case of poor recallers. In our dream groups we have found that even tiny fragments of dreams can yield valuable insights, sometimes life-changing insights, when there is nothing else to work on.

Anyone who is in a position of recalling one or more dreams every night may find it hard to work on them all in detail and will almost certainly use some commonsense selection process in deciding which ones to study. I do not see any objection to this, since in my experience the dreaming mind is very repetitious: if you fail to get a message from one dream, it will normally be thrown up again in another, especially if it concerns a really important problem in the dreamer's life. I would *not* suggest, however, that the prolific dream recaller should let his less interesting dreams or less coherent fragments go unrecorded, for I have found again and again that dreams shed light on one another over a period of time.

Sometimes several apparently quite different dreams of the same night or of consecutive nights will show the mind wrestling with the same problem, first in one way and then in another, trying out various solutions and forming further dreams on the apparent success or failure of preceding ones. On other occasions a dream theme that emerges as of major importance at one point in a person's life will be found, in retrospect, to have made its appearance earlier in less obvious forms, and this may serve to reveal what kinds of situations in the dreamer's life trigger this particular problem. Yet again, when dreams prove impossible to interpret on first studying them, a breakthrough of insight may come later, with many dreams and fragments over a long period suddenly coming together in a pattern, bringing a feeling of excitement like that which attends an important scientific discovery or the solution of a detective story.

And finally, there are dream fragments and short, boring dream episodes which later turn out to be extrasensory glimpses of a clairvoyant or precognitive nature. So I would urge all serious players of the dream game to keep some kind of diary in which to record *all* the dream material they manage to catch. For those with poor dream recall, the very fact of making the effort to write dreams down on a regular basis is itself a stimulus to making dreams more accessible, since it is a token to the dreaming mind of our willingness to take it seriously.

Keep some kind of recording system—pencil and paper or tape recorder—available by your bedside to be sure of capturing any dream you awaken from spontaneously, either in the night or morning. Dreams are fragile, and the brain appears to be in no condition to lay down memory traces when in the sleeping or semi-sleeping state. This is why the dreams we sleep through, or forget on waking, are usually gone forever, although occasionally dream fragments may return later in the day. The only sure way to catch a dream for future reference is to write it down or tape-record it immediately on waking, before it starts to fade. If you awaken in the night, it is no harder to get back to sleep after recording your dream than if you just turn over and close your eyes again. Those who think of themselves as poor recallers are often astonished to discover on starting the recording habit that they "have" far more dreams than they hitherto realized—and this discovery itself stimulates further improvement of recall, especially if it leads to some useful dream insights.

The following is a step-by-step procedure for collecting dreams and keeping a dream diary. It has been used with success by a great many people, but you may find after a time that you evolve your own method. In any case, follow the rules you yourself find most useful, and if you come up with any really original ideas of your own for catching dreams, please let me know!

INSTRUCTIONS FOR RECALLING AND RECORDING DREAMS

RULE 1: HAVE YOUR RECORDING EQUIPMENT HANDY.

If you are using pen and paper to record your dreams, be sure to have them within easy reach of your bed before you fall asleep. If you do not keep them handy, you will probably lose your dream, for even should you feel inclined to search for them during the night, your dream will probably have vanished by the time you find them.

If you sleep alone, you may prefer to use a tape recorder and transcribe your dream onto paper later. It is a weird and wonderful experience to play back your dreams the following day; your voice sounds as though it comes from another world—as indeed it does if you record while still half asleep. This is an excellent state in which to record dreams, an altered state of consciousness in which we almost touch the strange world of dreams, with its own unique laws and different space-time dimensions. I advise the tape recorder method of collecting dreams wherever possible.

If turning on the light during the night is likely to awaken your partner, then keep a flashlight by the bed and use this instead. An alternative is a Nightwriter pen of the kind used by policemen and nurses, which has its own light. It is also a good idea to keep extra pen and paper in the bathroom, as you can always write down your dream there if you have to visit during the night. Do not give yourself a tiny scrap of paper or a short length of recording tape—this is a vote of no confidence in the dreaming mind's ability to produce material for you, and your recall will suffer accordingly.

RULE 2: DATE YOUR PAPER OR RECORDING TAPE IN ADVANCE.

The date of a dream can be very important in understanding its meaning. It may turn out to be one of those "anniversary" dreams I mentioned in Chapter 1. One of John's dreams eluded interpretation for several months until a chance look at the date at the head of the paper—31st December—made him realize that his dreaming mind had been conducting a kind of review of the past year of his life, giving him a much more optimistic view of it than he had in his waking mind.

If you make a habit of dating your paper or recording tape each night before you go to sleep, it not only obviates the need to put a date on later, but also acts as another token to the dreaming mind of your readiness to await its messages. As a further token in the same direction, write down *Dream 1* after the date to show that you are ready for more than one dream in the night.

RULE 3: ENCOURAGE YOUR DREAMS BY MEANS OF SUGGESTION OR PRAYER.

When you are ready for sleep, turn off the light and relax in bed. Just lie back and let yourself sink right into the mattress. If your body is tense, deliberately relax each muscle in turn, starting with your toes and working all the way up the body until you reach your head. It is particularly good to relax the facial muscles, especially those around the mouth and jaw. This relaxation exercise is beneficial to sleep whether you are trying to collect dreams or not, but it also induces a "suggestible" state of mind in which you can influence yourself to remember your dreams.

When you are completely relaxed, say something like this: "Dreams, I'm ready for you. I promise that if you send me a message tonight, I'll do my best to remember it and write it down." Repeat this request several times and try to fall off to sleep with it in mind.

Religious people may prefer to use a prayer to God for guidance instead of a request to "dreams." After all, it is a major part of all religious traditions that God speaks to man in dreams.

If you prefer a firmer approach, or if you find that suggestion or prayer does not bring results, try the following autohypnotic procedure. Relax as described above and send yourself to sleep repeating the words "I shall awaken with a dream in mind, and I shall write it down." If you do not wish to awaken during the night, say, "I shall awaken with a dream in mind *in the morning,* and I shall write it down." The dream game should not be a chore, and if you do not normally wake during the night there is no need to force yourself to change your habits. The last dream before waking in the morning is usually long and involved, and if you can train yourself to catch that fairly regularly, you will have ample material for dream work.

RULE 4: NEVER TRUST YOUR UNAIDED MEMORY AND BEWARE OF PROCRASTINATION.

Because the memory traces of dreams fade so quickly, never deceive yourself that you will remember a dream without recording it immediately just because it seems clear to your mind at the moment. During my years of experimental dream research, I studied the problem intensively, and found that when my subjects gave me a dream in the middle of the night, as much as fifty percent of the information I had recorded then was forgotten by morning, in spite of the fact that they had been forced to make the effort to relate the dreams to me on waking from them.

Similarly, research has shown that however vivid a dream may be in your mind when you wake in the morning, you will have forgotten a good deal of it if you rush to wash or have breakfast before making a record of it, still more if you go to work and leave it until later in the day.

Procrastination is the thief of dreams and can be the thief of sleep into the bargain. In my early days of dream collecting, I would often find myself lying awake with a dream in mind, knowing that it was important, but not quite able to muster the willpower to write it down. Torn between the need to hang onto the dream and the desire to fall asleep, I would often stay awake for an hour or more turning it over in my mind, until I realized that I lost far less sleep by making the effort to record the dream immediately on waking.

RULE 5: NEVER DISMISS A DREAM AS TOO TRIVIAL TO RECORD.

Do not have any preconceived ideas about what a dream is. If you awaken with *anything at all* in mind, write it down. Even thoughts and impressions can be interesting, telling us how we felt about some event of the day, at other times being precognitive in nature. Jung put it very nicely when he wrote, "The dream is often occupied with apparently very silly details, thus producing an impression of absurdity, or else it is on the surface so unintelligible as to leave us thoroughly bewildered. Hence we always have to overcome a certain resistance before we can seriously set about disentangling the intricate web through patient work. But when at last we penetrate to its real meaning, we find ourselves deep in the dreamer's secrets and discover with astonishment that an apparently quite senseless dream is in the highest degree significant, and that in reality it speaks only of important and serious matters."

If a dream eludes you, don't worry. Strain does no good either to you or your dream recall. If the message is important, your dreaming mind will bring it up again later on.

RULE 6: RECORD YOUR DREAM AS FULLY AS POSSIBLE.

Record as much of your dream as you can remember, either fully or in note form. You may like to lie relaxed in bed for a moment or two, turning the dream over in your mind before you actually record it, for there is some evidence that a sharp body movement may cause it to disappear. If you do this, resist the temptation to drift off to sleep again; maintain an attitude of relaxed attentiveness and you will find that the dream scenes gradually come back to you. Having got back as much as you can, write it all down or tape-record it. If you write it in the present tense, it may help to keep the dream alive.

Add any immediate associations or ideas about the possible meaning of the dream and write down any feelings connected with the dream. Did you feel happy, depressed, afraid, excited, or neutral? This can make all the difference to an interpretation.

Note also your mood on waking. Do you feel good or bad, happy or sad, peaceful or worried? This is important, for if you feel worried after an apparently good dream, there is something that has eluded you and you will need to look for this when you come to work on the dream. The same thing applies to an apparently bad dream that leaves you feeling unaccountably good—so be sure to make a note of your feelings in the dream itself and your mood on waking.

Write down also any color that stands out in a dream, however silly or obvious it may seem, as the dreaming mind often uses the device of painting an image or scene in vivid color in order to draw your attention to its significance. For example, my husband once dreamed of Queen Victoria poking his nose with an enormous acupuncture needle, and on relating the dream he remarked how red the blood was. The redness of the blood was, in fact, the vital clue to the meaning of the dream, which was pointing out his feeling of inferiority at possessing very ordinary red blood, instead of the blue blood others of aristocratic birth had flowing in their veins!

Actually, we all dream in color most of the time, as dream researchers have demonstrated by waking subjects from REM sleep and questioning them immediately about color in their dreams. Subjects almost always replied that the grass was green, the sky blue, the house white, or whatever. However, the memory for color in a dream disappears very rapidly, as indeed it does in waking life, unless something particularly catches the eye. So if your attention is drawn to a specific color in a dream, always make a note of it at the time of recording.

RULE 7: (FOR REAL ENTHUSIASTS!)

When you have recorded a dream during the course of the night, do not go back to sleep until you have invited the next one to come to you, by writing beneath your record or speaking into the recorder the words *Dream 2.* If you collect a second dream during the night, end your record by adding the words *Dream 3.* Sometimes you may collect three or four dreams during the course of a single night's sleep. This is always an interesting exercise, as dreams of the same night usually relate to the same problem or situation in your present life, and one dream is often useful in throwing light on the meaning of another.

Once again, do not strain to catch your dreams, and allow yourself sufficient sleeping time if you try this experiment, perhaps by going to bed a little earlier than usual.

RULE 8: TRANSCRIBE YOUR DREAMS THE FOLLOWING DAY.

Try to make time to transcribe your dream notes the following day, or as soon as possible afterward. Allow a separate sheet of paper or card (5″ by 8″ is a convenient size) for each dream, which should be clearly dated. If you have more than one dream in a single night, be sure to number them in order of appearance—Dream 1, Dream 2, Dream 3, and so on. Add any further ideas you may have about a dream while transcribing it, but avoid forcing or elaborating the dream material. File your dream records in order in a loose-leaf folder or filing box.

RULE 9: RELATE THE DREAM TO AN EVENT OF THE DAY.

When you have transcribed your dream, together with all associations and feelings, try to relate it to an event of the previous day, as this gives the vital clue to the meaning of the dream. For example, if you are angry or afraid in a dream, ask yourself whether or not any event of the day could have sparked your anger or fear, perhaps without your consciously realizing it at the time. Sometimes the dream makes it easy for us by depicting specific people, places, or events, but more often the dream uses symbolic imagery, and it may appear at first glance to be totally unrelated to anything you can think of. If you are unable to relate your dream immediately to anything that happened or was on your mind the day before, then jot down the day's major landmarks, so that when you come to work on the dream later you will have at least one or two points of reference from which to work. In particular, make a note of any TV program you were watching, or book you were reading, or conversation you were having just before going to bed.

A cautionary note here: never be tempted to think that the TV program, the book, or the late-night conversation can "explain away" your dream. Such events are merely triggers for a dream fantasy of your own making, and obviously a hundred people watching the same TV program will have a hundred different dreams about it. Similarly, if you awaken from a dream of fire-engines to find the alarm clock ringing, or from a dream of being crushed to death by a bear to find a heavy blanket on top of you, by all means make a note of these things, but remember that they are insufficient to explain your own particular dream fantasy; after all, another person subjected to exactly the same stimuli may dream of church bells or a lover's embrace. The same applies to internal body stimuli, such as bladder pressure, thirst, or indigestion; all these undoubtedly influence dreams but there are an infinite number of different dream stories that can be woven around them. Both external and internal stimuli may be picked up by the dreaming mind and incorporated into *an already ongoing* dream, but they cannot in any way initiate a REM period before its time or "cause" you to dream.

Incidentally, you do not get the best out of the dream game if your sleep is disturbed by indigestion or intoxication. It used to be thought that cheese and pickles eaten late at night increased our quota of dreams, but modern research shows that such pressure on the digestive system simply makes us more restless, causing us to wake up more often during the night with vague, rambling dream fragments in mind (mostly from non-REM sleep) which are usually not very useful for subsequent interpretation.

I am often asked whether, as I consider dreams to be so important, I would recommend the use of an alarm clock to "force" awakenings during the night, as a rough-and-ready, do-it-yourself counterpart of what experimental dream researchers do in the laboratory. My answer is that under normal circumstances such drastic measures should not be necessary, although the alarm clock experiment is an interesting one for real enthusiasts to try once in a while. The idea is to set the clock to awaken you every two hours or so throughout the night, and then record whatever is passing through your mind on waking. You are almost certain to catch at least two REM dreams, and you may also catch some non-REM dream material, which would then enable you to detect the difference between the two types of dreaming. I suggest separate rooms for married players while the alarm clock experiment is in progress!

YOUR DREAM DIARY

The rules I have given represent an idealized approach to keeping a dream diary, and I have spelled it out in detail so that you can refer back to it from time to time if your own experiment fails to come up to expectation. It may be that you are missing out on one or two small points, which can be rectified by rereading the chapter. In my experience, it is always the two most important rules that are ignored, namely to write down the dream on waking, and then to relate it to some event of the previous day or two. Failure to carry out the first rule results in loss of dreams, while failure to keep the second renders effective interpretation almost impossible, as later chapters will show.

At this point, if you have been following the instructions as you read this book, you should already have the beginnings of your dream diary. The most exciting part, however, is yet to come—the meaning of your dreams, revealing the secrets of the hidden life you live during the dark hours. I have suggested transcribing your dreams into a loose-leaf book so you can add your interpretations later, together with any further thoughts you have about the dreams and their symbols. As Edgar Cayce's son, Hugh Lynn, said recently at Virginia Beach, the best book on dreams you will ever read is the one you write yourself.

PART II

PLAYING THE GAME

«4»

THE LANGUAGE OF DREAMS—OR WHY THE UNIVERSAL LANGUAGE HAS NO DICTIONARY

Even a relatively stupid feeling out of one's own images is better than reason or the guess of the best outside experts. Outsiders are inclined to project their own lives into one's own images. Occasionally I have seen very good friends make meaningful guesses about another's images. But you are the life that projects your images. Your most halting understanding is closer to its own source.

—Wilson Van Dusen

Some dreams are very simple to understand and require no interpretation. If you dream of having sexual relations with your reasonably attractive neighbor or colleague, your dream is probably doing no more than expressing an erotic attraction toward him—and this may come as a real revelation if you are not aware of these feelings in waking life. Similarly, if you dream of killing someone in your life at the present time, there is no problem in seeing that you feel some hostility toward him, even though you may be reluctant to confront these feelings consciously. In such cases, the dream is translating a thought of the heart directly and openly into picture language, and the very fact of actually experiencing this feeling in the dream makes it easier to accept and confront in waking life, which we must do if it is not to ruin our lives by festering away underneath and eventually finding unconscious expression in some devious way.

Sometimes a dream will give a dramatized picture not only of a hidden feeling about ourselves or others in our life, but also of the reason for that feeling. For example, many years ago while still married to my first husband, I dreamed of creeping up on him while he slept and bashing him to death with a carpet sweeper. The dream expressed hostility toward him in no uncertain terms, but why did my dream-self kill him with a carpet sweeper, rather than with a golf club or hammer? Only a very little imagination is necessary to hazard a guess that my resentment had something to do with domesticity, though the dream was not specific about the actual issue. My husband was, in fact, insisting that I relinquish my career in order to look after the home and family, a prospect that was evidently making me feel murderous toward him, although I had not consciously faced the full extent of my hostility. The dream therefore warned me in no uncertain terms that if I acceded to his demand I would become so frustrated and unhappy that I might actually be driven to harm him in some way, perhaps without realizing what I was doing. Had I been an artist painting a picture of the situation, I could not have depicted the conflict in a clearer fashion. Had my resentment been triggered by the long hours he spent at his workbench or on the golf course, my dream might have shown my killing him with a hammer or golf club.

Even the most involved, bizarre dreams are performing essentially the same function as the more obvious ones, namely expressing feelings, emotions, and intuitions in picture language, but their pictures are more complex because they are expressing more complex thoughts. For example, had the thoughts of my heart included the feeling that my husband was a lousy male chauvinist pig, then my dream might have depicted my creeping up on a sleeping pig and watching it turn into a louse as I killed it. And had the scene been set in my husband's childhood home, then my dream might be expressing the feeling that his upbringing was responsible for his unreasonable behavior. On the other hand, if the dream-murder took place in my own childhood home, or was followed by my husband turning into my mother, then the dream could be relating the present situation to a similar feeling of resentment experienced in the past, perhaps when my mother made me do the housework while my brother went out to play. Dreams often condense several different ideas and impressions into a single compact image in this way.

THE LOGIC OF THE HEART

Why does the dreaming mind use such complicated forms of expression? Freud, who pioneered the modern understanding of dreams, believed that it expressed itself in symbols in order to disguise from the dreamer the full unpleasantness of all his primitive sexual and aggressive drives which are repressed during waking life because they are not considered "nice" or proper. The obvious time for them to find release, said Freud, is during sleep, when the conscious mind is off-duty and they can express themselves in fantasy without causing any actual harm. The dream resorts to symbolic, disguised picture language, he held, for if it were to depict sexual and aggressive acts openly the dreamer would awaken in horror and lose his much-needed sleep. Thus, the dreaming mind would cunningly show the dreamer breaking his father's umbrella and riding on the back of a cow, allowing him to sleep on undisturbed by any awareness that he had just indulged the fantasy of castrating his father and having sex with his mother.

This theory has had a pernicious influence in spreading the idea that dreams are out to deceive us with disguises that can be penetrated only by an expert who will force the dreamer to accept the unpleasant truths he is trying to avoid. Even in Freud's own day some of his followers, of whom Jung was the most notable, objected that concealment is essentially a function of the waking mind, quite out of place in considering the mind's deeper processes, and this is borne out by modern experimental dream research which has given us access to a wealth of information about dreams and dreaming that was simply not available to Freud. We owe him a permanent debt for persuading the educated world to recognize that dreams are not nonsense, and for giving us many valuable insights about the way in which the dreaming mind uses symbolic picture language to express unconscious mental processes. But we now know that his "disguise" theory is an unnecessary complication based on his obsession that the unconscious is composed wholly of unpleasant drives which the conscious mind cannot face.*

It is now widely accepted that symbols in dreams reveal rather than conceal the truth from the dreamer, presenting it, moreover, in a very precise, condensed form. For example, if I dream of killing a pig one night, and the next night dream of bashing my husband to death, then it is more reasonable to suppose that the pig symbol represents additional feelings about my husband's male chauvinism, rather than serving as a (most inefficient) disguise to spare me a painful confrontation with my murderous feelings. In fact, the basic reason for doubting Freud's disguise theory is that the pictures thrown up in dreams are no different from the metaphorical and slang expressions we use all the time in waking life. Far from being

*For a full discussion of Freud's theory see *Dream Power.*

a disguise, it seems that symbolic language is quite the most efficient way of articulating a whole constellation of feelings, and this is true whether the feelings are nasty or noble.

Poets, playwrights, movie makers, and artists have always used this language. In many ways a dream is very like a movie that flashes a series of pictures across the mind's eye during sleep, conveying its message by means of visual imagery and association of ideas. And although this picture language often seems strange and complicated at first, we find as we come to understand it that it is actually the clearest, most economical of all languages for expressing the subtleties and intricacies of human experience.

Thinking by means of pictures and pictorial idea association is probably the most primitive of all modes of thought, going right back to the dawn of the human race when the power of speech was only just beginning to develop, and abstract thinking had not yet been born. The caveman glancing across the cave at his woman probably actually *saw* her as a she-wolf or a deer long before he was able to articulate the concept that she was *like* an animal—a form of thought that may lack scientific accuracy but has great emotional immediacy. In fact, we still rely on it for expressing feelings even though our species has now developed language and the power of abstract thought, for the bulk of our speech consists of pictures translated into verbal metaphors—"gay dog," "wolf in sheep's clothing," "ships that pass in the night," "bridge over troubled waters," "climbing to the top of the tree," "missing the bus," "getting into hot water," "going through the roof," and countless comparable turns of phrase on which every language on earth relies for its emotional impact and meaning in all but the most impersonal scientific matters. So when the dreaming mind expresses itself in movie terms, cutting out all the "as ifs" and showing us literally crossing roads and bridges when we are facing major life decisions, or literally being devoured when we feel "eaten up" by something, it is using the most fundamental of all languages, shared by men and women of every age and race.

Primitive man may possibly have had dreams that were purely pictorial, somewhat like silent movies, but this is something about which we can only speculate. Ever since words became an essential part of human life, dreams have used them, but their role is usually subordinate to the expression of feeling contained in the picture story. For what the dreaming mind is after is the total expression of some feeling or intuition in all its complexity, subtlety, and depth, and action pictures are far more expressive and less restrictive in this respect than words. Of course, we must translate our dreams into the language of the head if we are to get ourselves together in some tangible way—but even after the most meaningful interpretation, the dream in its totality still remains a *feeling* statement of a different order which appeals to the heart in much the same way as a painting or a poem. So while in one sense the picture language of dreams is more primitive than the language of waking life, this does not mean that it is in any way crude, incoherent, or illogical. The dreaming mind uses this language for a purpose—to express the thoughts of the heart in the fullest possible way—and it does it so richly and subtly that many of the ancients called it the language of God.

The language of dreams has, in fact, a logic of its own which, while often wildly different from the mechanical logic of our head thoughts, is actually quite exact in relation to the thoughts of the heart, and is, moreover, perfectly familiar to us when we take the trouble to understand it. Thus, in a dream it does not seem in the least surprising that a human being should turn into an animal, or that I should be carrying out some activity in my childhood nursery, for the simple reason that during the day my heart has already *seen* that person as an animal or picked up my memories of childhood. In the same way, it is a common feature of dreams that you have only to fear something—that there may be a dangerous animal in a dark basement, or that a high building may collapse, or that a stray insect may herald a plague of them—for it to happen immediately, and this too seems perfectly natural in a dream, for in the language of the heart, to fear something *is* to imagine it actually happening. Again, the dreaming self does not bat an eyelid at defying the laws of nature by flying under its own power, or breaking the rules of social expediency by arriving at the office in pajamas, for the heart knows very well what it feels like to "fly high" or to "sleep on the job." When we are surprised by a dream, it is because the heart has itself been surprised in some way during the day, just as a dream of chaotic and confused events is an indication that the heart itself has been feeling muddled and not, as many people think, because there is anything inherently confused about the dreaming process itself. At the opposite end of the scale is that very special dream state in which we become aware of something incongruous with waking experience and realize in the dream that we are dreaming; this "lucidity," as it is technically called, is a higher state of consciousness reflecting an actual coming together of head and heart somewhere in waking life during the course of the day.

UNDERSTANDING DREAM SYMBOLISM

Calvin Hall, who restated some of Freud's important insights about dream symbolism in truly scientific, down-to-earth, and commonsense terms, summarizes the matter succinctly in his book *The Meaning of Dreams*: "There are symbols in dreams for the same reason that there are figures of speech in poetry and slang in everyday life. Man wants to express his thoughts as clearly as possible in objective terms. He wants to convey meaning with precision and economy. He wants to clothe his conceptions in the most appropriate garments. And perhaps, although of this we are not too certain, he wants to garnish his ideas with beauty and taste. For these reasons, the language of sleep uses symbols." He adds that "anyone who can look at a picture and say what it means, ought to be able to look at his dreams and say what they mean."

Now John, my present husband, says he can often stand in front of a picture for hours and still not know what it means. I sympathize with him, for I have felt the same way. While I go along with Hall in believing that it is usually possible for most people to agree on certain aspects of a painting—whether it expresses despair or joy, stillness or excitement, foreboding or hope, death or resurrection—I agree with John that we cannot possibly know for certain why the artist painted the moon in crescent rather than full or why that fish is flying through the air with a dog between its teeth. We can hazard as many guesses as we like, but in the end only the artist himself knows the meaning of the images in his painting. (If the artist says he doesn't know, then we have to say his picture has no single "true" meaning—it is like an inkblot test, something each person can interpret as he likes. But such cases are rare.) For this reason, I would change Hall's statement to, "Since anyone can (normally) look at a picture *he himself* has painted and say what it means, he ought to be able to look at his dreams and say what they mean."

For we, and only we, are the painters of our dream pictures, and learning to understand our dreams is a matter of learning to understand our heart's language—which is what this book is all about. A psychotherapist or dream guru may come along and suggest all manner of interpretations, but unless they resonate with your own bones and move you to change your life in some constructive way, then they remain useless speculations on his part, which logically is the same thing as saying they are just plain wrong. Moreover, I can guarantee to find you at least another dozen "experts" who would give you quite different "correct interpretations"—as happens to me every time I relate some of my own dreams at professional gatherings.

Dream books in which you look up the meanings of dream themes and symbols are equally useless, whether they be traditional or based on some modern psychological theory. While a dream of a banana may indeed be a warning not to eat too much to the compiler of one particular dream book, it is more than likely to be a phallic symbol for some other author. It certainly does not mean either to me, and is unlikely to do so for anyone else. You cannot learn dream language as you can learn French or Spanish, for each of us has his own individual dream vocabulary formed from his own unique life experiences. In Chapter 7 and the Appendix I give instructions for compiling your own dream glossary, but this cannot be used to interpret the dreams of your friends or even your partner, though sometimes a person will "lift" a certain symbol and its meaning from someone else if his dreaming mind considers it useful and relevant at the time. This explains to some extent why psychotherapists and other dream gurus manage to hold on to dogmatic theories about the meaning of dream symbols, for their patients and pupils often learn the master's dream language and incorporate his symbols into their own dreams. Similarly, people brought up in a close community, say in a primitive tribe or on an island where there is a strong tradition about dream symbols, will tend to dream in accordance with that tradition simply because it has been built into their education. But even such self-contained situations as these are not proof against individual variations, and the general principle is that each person's dream glossary is his and his alone.

Since we are all human, there are certain general *themes*

that come up from time to time in almost everybody's dreams, and these seem to have a similar meaning for all of us, simply because they are picture metaphors for a common kind of feeling or experience. In the next chapter, I discuss eight such common dream themes—falling, flying, appearing nude or scantily clad in public, taking an examination, losing teeth, losing valuables, finding valuables, and sex—as an introduction to the way the dreaming mind uses picture metaphors. (Other common dream themes such as paralysis, being pursued, forgetting lines, being burgled, arriving late, missing a boat, meeting famous people, and so on appear elsewhere in the book.) Even such common dream themes as these, however, cannot be called truly universal metaphors, for though they normally point to similar kinds of feelings or situations, they carry quite different specific meanings for each individual depending on his life circumstances at the time of the dream.

In Chapter 6 I describe the uncanny way in which dreams pick up and make use of puns, slang expressions, body language, and all kinds of colloquialisms in order to give a good, vivid picture of the dreamer's feelings. This is one of the most important aspects of the language of dreams, and not enough has been made of it. Here again, however, only the dreamer himself knows the specific life situation to which his dream refers when it shows him "missing the bus," "going under the table," "pulling up his socks," or "getting too big for his boots."

When it comes to understanding the individual symbols in your dreams, such as animals, vehicles, buildings, mountains, rivers, trees, and so on, it is almost impossible to make generalizations because our individual experiences are so different. Freud tried to construct a theory of universal dream symbolism based on the notion that dreams originate mainly from a layer of the mind formed in early childhood when all human beings are preoccupied with the same basic experiences of the body and its functions and learning to relate to parents and siblings. So we have the well-known tendency of psychoanalysts to interpret long, thin objects as penis symbols, openings as anus or vagina symbols, closed spaces as womb symbols, male characters and fierce animals as father symbols, female creatures and soft animals as mother symbols, insects and small animals as siblings, rounded fruits as breast symbols, and so on. Today, when modern research has shattered the notion that dreams derive primarily from childhood experience, Freud's whole theory of universal symbolism falls apart. At best, his ideas are suggestions about what *some* dream symbols might *sometimes* mean; at worst, they can be positively misleading, by biasing people to look for only one kind of meaning in their dreams, when the very first rule should be to keep on the alert for all possible meanings a dream can have in the light of your own experience at the time.

In Chapter 7, I give instructions for discovering the meanings of your own dream symbols and the relevance they have to your current life situation. For example, if a snake appears in your dream, the first question to ask (which most dream books and dream gurus tend to ignore altogether) is whether or not there are any actual snakes in your life at the present time. The dream could be a warning about a snake you saw out of the corner of your eye in the bushes in the corner of the yard the previous day but were too busy to register consciously at the time—as actually happened to me in North Carolina. Such a warning could save a life. But even where there is no possibility of real snakes, and the dream has to be understood symbolically, the only general statement that can be made about the dream snake is that it must be a symbol for something or someone in your life. Precisely what it stands for will depend not only on your specific circumstances, but on what the idea of snake means to you. If it conjures up a picture of a slimy, poisonous creature, then the dream is expressing your feeling that someone in your life is a slimy, poisonous individual, and this could be another person or it could be a part of yourself. If you like snakes, on the other hand, then the dream represents a beneficent force in your life, and the difference is obviously crucial.

In my experience of many thousands of dreams, cases where snakes turn out to be sex symbols, as in most psychoanalytic interpretations, are quite rare. People whose minds have been strongly influenced by a biblical upbringing often find that dream snakes symbolize some kind of temptation, but not necessarily a sexual one. Edgar Cayce commonly interpreted snakes as temptation, and I'm sure this holds good for many of his followers who have absorbed the association into their thinking—but he himself would have been the first to insist that there is no universal necessity for it to be so and would have urged individuals to check whether or not some other association to "snake" could be operative for them in any particular dream. Among the Cayman Islanders, for example, there is a strong tradition that dream snakes indicate that you have secret enemies, which means of course that they are likely to have such dreams whenever they have picked up subliminal perceptions of treachery during the day without consciously registering cause for concern. Colloquial expressions like "snake in the grass" come to mind here, and this seems to be quite the commonest meaning of dream snakes shared by many different people the world over who have a horror of snakes, even though they may never have seen a real one.

In my own dreams, snakes appear to suit the occasion; for example, whenever I have been indulging rather heavily in spiritual disciplines like meditation or yoga, my heart produces a bevy of dream snakes as a reminder for me to come down to earth and pay attention to my physical life—and my most recent snake dream occurred on the first night after our arrival on a small island. I dreamed that it was overrun by friendly snakes, and on waking fully expected to find that I had picked up subliminal perceptions of snake infestation on our evening walk the night before. I was almost disappointed to learn that the island boasted only two species of very harmless grass snake. As the snakes in my dream were friendly, it seemed less than civil to interpret the dream in terms of secret enemies, and the meaning did not hit me until I recalled that I had been reading about the ancient Greek practice of dream incubation the previous day, in which patients would sleep in the temple of Asclepius hoping for a dream cure from the God.* Apparently harmless yellow snakes would crawl over the sleeping bodies in the temple—which resonated immediately with my heart's feeling that this island would be the perfect place to open my own dream center for healing and regeneration through modern dream therapy.

Snakes were common symbols of medical art in the literature of the ancient Greco-Roman world, so it is not surprising to find reports of people dreaming of snakes when they were seeking or undergoing any kind of therapy in classical or medieval times. In fact, many societies all over the world have used snakes as symbols of healing wisdom, and it is this kind of phenomenon that has led several psychological writers to speculate that certain symbols are virtually built into every human mind irrespective of individual life circumstances. Jung, who objected strongly to Freud's belief that all dreams derive ultimately from the experience of early childhood, was nevertheless a firm advocate of the notion that there are certain universal patterns in the unconscious which have a tendency to emerge in dreams, works of art, legends, fairy tales, and myths in all human societies. He believed these universal

*C. A. Meier, *Ancient Incubation and Modern Psychotherapy* (Evanston, Northwestern University Press, 1967).

patterns formed a "collective unconscious" common to all mankind and speculated that they might actually be inherited in the brain structure. For him the snake was just such an "archetypal" symbol, expressing man's deepest awareness of the life-energy of nature, of which sex could be seen as one special aspect, temptation another, and healing another, while the most profound of all would be the ancient symbol of the snake swallowing its own tail, the *Ouroboros,* which expresses the way the life-force in nature feeds on itself and renews itself eternally.

The idea of archetypal symbolism has fascinated a great many people, mainly, I suspect, because it adds a dimension of almost metaphysical significance to the workings of the human soul, and I empathize with this feeling; but for the practical purpose of understanding dreams the theory has precisely the same drawbacks as Freud's more earthy notion of universal body symbolism. There may indeed be a level at which our minds and hearts wrestle with universal problems of metaphysical dimensions—I believe there is, as later chapters will show—but it is certainly not the case that all dreams arise from this level, as Jung himself continually emphasized in his writings and all sensible Jungian therapists know very well in their practice. Hence, in any particular case the only way to find out what a dream image means is to discover what the dreamer himself thinks it could stand for in his own mind at the time, and any "archetypal" meaning distilled from the world's literature is at best a suggestion to be made as a last resort when all personal associations fail. All too often, however, people who have read Jung, or got into the hands of the wrong kind of guru, go searching for archetypal meanings in every dream and thereby miss simple meanings, losing themselves in a sea of metaphysical abstraction instead of coming to terms with down-to-earth problems.

Even when you are dealing with one of those rare dreams

that do seem to come from the deepest, most transcendental levels of the soul, the idea of archetypal symbols and universal meanings can still be thoroughly misleading, by failing to take account of the vast differences between the ways people experience life in different parts of the world. For instance, in Western literature and in most Western peoples' dreams, the sun is a symbol of life and warmth and hope, but a Saudi Arabian tells me that in his dreams and those of his friends, the sun stands for destruction and desolation, reflecting the daily experience of the sun in the lives of people in scorching desert lands. Of course, both life-giving warmth and scorching destruction can be interpreted as experiences of energy (we have in Western mythology the story of Icarus being scorched by flying too close to the sun) but at this point the "universal" meaning of the symbol becomes so broad as to be quite useless for all practical purposes. Confronted with a dream of the sun, we simply come back to the need to find out what the sun means to the particular dreamer in his present life. If he resonates to some suggestion you make as a result of your reading books on sun symbolism in world literature, all well and good, but this proves nothing about any inbuilt meaning of sun symbols in the human mind.

So while it is true that the dreaming mind speaks a picture language that is common to all human beings irrespective of race, creed, sex, or the age in which they live, it is a language that has no dictionary except the one each dreamer compiles for himself, and even that dictionary is not totally fixed, since the same person can use symbols in different ways in different circumstances. Moreover, the same dream can sometimes carry more than one meaning at different levels, as I shall show in later chapters. However, the art of learning to understand dream symbols is basically quite easy once you get started, and in the following chapters I shall be taking you through some simple exercises to show you how it goes.

A final word of caution about dreams in general: remember that they are always focused on your own personal life, hopes, and fears. It is very rare indeed for people to dream about general subjects like politics, science, social affairs, or religion, except insofar as the dreamer is personally involved with such matters in the intimacies of his own life. As Calvin Hall says, "A dream is a personal document, a letter to oneself. It is not a newspaper story or a magazine article." Sometimes we do have dreams with a moral for society as a whole, but these are rare, and in any case they arise out of our own needs at the time of the dream. Even Jung, the archexponent of the mind's universal dimensions, insisted that "one should never forget that one dreams in the first place, and almost to the exclusion of all else, of oneself."

«5»

EIGHT COMMON DREAM THEMES

Those things that have occupied a man's thoughts and affections while awake recur to his imagination while asleep.

—Thomas Aquinas

In this chapter, I shall discuss eight common dream themes that crop up time and again in lectures, correspondence, TV and radio shows, and even in casual conversation—namely, falling, flying, nudity, taking an examination, losing teeth, losing valuables, finding valuables, and sex. Statistical surveys have shown them to be among the commonest dream topics in Western civilization, and although there are many others equally common (which I discuss elsewhere in the book), I have chosen these eight themes to illustrate how dream interpretation works. Novices at the dream game who have collected a few recent dreams may be able to make a start at understanding them by following the basic principles laid down in this chapter, and even old hands and professionals should find some useful hints here, since my own empirical approach does not depend on any theory of dream symbolism, as most schools of dream interpretation have done in the past.

The basic rules of the dream game at this stage are:

1. The dream should always be considered literally in the first instance and examined for signs of objective truth, such

as warnings or reminders, before moving on to metaphorical interpretation.

2. If the dream makes no sense when taken literally, then (and only then) should it be seen as a metaphorical statement of the dreamer's feelings at the time of the dream.

3. All dreams are triggered by something on our minds or in our hearts, so the primary objective must be to relate the dream theme to some event or preoccupation of the previous day or two.

4. The feeling tone of the dream usually gives a clue as to what this particular life situation is. For example, if the feeling tone of the dream is miserable, then the dream was sparked by some miserable situation in the dreamer's current life.

5. Common dream themes like those discussed in this chapter are likely to indicate common areas of human feeling or experience, but within these broad limits each theme can mean quite different things to different dreamers according to the individual's life circumstances at the time of the dream.

6. The same dream theme may recur from time to time in the dreams of the same dreamer and have a different specific meaning each time, according to his life circumstances at the time of each dream.

7. Dreams do not come to tell us what we know already (unless of course, it is something we know but have failed to act upon, in which case they will recur, often in the form of nightmares, until we do), so if a dream seems to be dealing with something you are quite well aware of, look for some other meaning in it.

8. A dream is correctly interpreted when and only when it makes sense to the dreamer in terms of his present life situation and moves him to change his life constructively.

9. A dream is incorrectly interpreted if the interpretation leaves the dreamer unmoved and disappointed. Dreams come to expand, not to diminish us.

The following examples of common dream themes aim to show how to put these principles into practice in understanding your own dreams, by giving the feel of how dreams work without forcing any particular interpretations. My discussion by no means exhausts the host of possible alternate interpretations you may find to your own dreams—in fact, the whole point of this chapter is to stimulate you to look for your own meanings in the dreams you have collected, in the light of your own current life experience.

DREAMS OF FALLING

The first step is to see whether or not the dream contains some kind of warning of a possible literal fall in your life because of something you have been neglecting or have failed to register consciously during the previous day or two. For example, when I dreamed of falling off the balcony of our new seventh-story apartment, I immediately examined the guardrails on waking and found them distinctly rickety. This information had obviously registered at the back of my head the previous day, but I had been too preoccupied to take conscious note of the potentially dangerous situation. The same rule applies when we dream of someone else falling. For example, our neighbor dreamed that his son fell off a ladder, and having talked with us about taking dreams literally in the first instance, he examined their ladder and discovered a loose rung. Once again, the watchdog of the psyche helped prevent a nasty accident.

However, if a falling dream carries no literal warning message of this kind, the next step is to ask what kind of metaphorical "fall" the dreamer could currently be concerned about. Looking through my dream collection, I find the following examples: a colleague who dreamed of falling down the college stairs at a time when he feared demotion on account of poor work, reflecting his fear of loss of status; a schoolboy who dreamed of falling down the school stairs the night after he presented his parents with a poor report card, reflecting his feeling that he would now fall in their estimation; the wife of a radio station director who reported a series of falling dreams soon after her husband had been promoted, reflecting her feeling that she could not keep *up* with him; a teenage girl of a Catholic background who related a series of very unpleasant falling dreams soon after she started sleeping with her boyfriend, reflecting deep guilt feelings about her "fall from grace and virtue" in the eyes of God.

In all these dreams, the falls were experienced as unpleasant, but this is not always the case, in which event the dream is showing that some "fall" we have experienced, or are anticipating in waking life, is seen by our hearts as less fearsome than perhaps we think, or maybe even downright pleasant. For example, when I appeared on the TV program *To Tell the Truth,* Peggy Cass, the actress, related a dream she had some years previously in which she fell from the top of the Empire State Building onto a beautiful soft bed of pine needles. Having ascertained that the falling sensation was pleasurable, and bearing in mind that she landed on a *bed,* I suggested that she saw herself as a "fallen woman." Amid the laughter, she was heard to protest, "Not any more . . ." This dream disposes of the widespread myth that you will die if you hit the bottom in a falling dream; the whole idea is nonsense, for quite apart from the fact that thousands of people who have hit bottom in a dream, including Peggy and myself, are still alive and well, how can we ask those who died in bed whether or not they had a falling dream at the time of death?

DREAMS OF FLYING

Since dreams of flying under one's own power cannot have any literal significance (except perhaps for astronauts), we have to look for their meaning by converting the picture into an idea or thought—that is, they express the dreamer's feeling of being "high" or "on top of the world" in his life at the present time, or perhaps his struggle to "rise above" circumstances or avoid restrictions. Once again, the feeling tone—elation or anxiety—gives the clue to how you really feel about these events in your life. For example, Johnny Carson, host of the *Tonight* show, often dreams of flying and performing acrobatics in the air after a good day in the studio. He thoroughly enjoys these dreams, which indicates that he really "gets a lift" out of showing off his verbal acrobatic skill. And I myself flew for days in my dreams after the publication of *Dream Power,* which came as quite a revelation as to how very delighted I felt to become an author.

Some years ago, I experienced a series of flying dreams in which I was trying to escape from a threatening situation. I was pursued by an enormous green, shapeless monster, and I would usually awaken in terror as the thing enveloped me. The dreams occurred at a time in my life when I was battling jealous feelings about my first husband, so I had no difficulty in identifying the "green-eyed monster" of my nightmares, and the message was clear. Whereas my head was quite convinced that I had these feelings under control, the thoughts of my heart showed quite clearly that my efforts to "rise above" them had failed and that I was still "consumed" with jealousy. The dreams continued, with variations according to the events of the previous day, until I faced the unpleasant truth that I was not as uninvolved as I believed and discussed the subject openly with my husband. Only then was I able to get myself together and view the situation rationally.

The height to which you fly in dreams is important, as well as the feeling tone. Jimmy Dean, the comedian, told me a dream in which he was flying along at medium height, having discovered that he felt distinctly anxious if he ventured higher. When I asked how he felt about "flying high" in his life, he replied, "Funny you should ask that. I've just been offered my own show on Broadway but turned it down because I didn't feel quite *up* to it." His heart and head were obviously together on the subject of ambition.

Psychoanalysts tend to reduce all pleasant flying dreams to sexual desire, but this is at best a gross overgeneralization and can be a put-down of one of the most extraordinary phenomena of dreams. My research has shown that when flying dreams have erotic overtones they are usually forerunners of an out-of-the-body experience. I have found that very often a flying dream is initiated by sexual energy circulating in the region of the lumbar plexus at the base of the spine or around the sexual organs. If the energy manages to flow up the spine to the top of the head, the dreamer takes off into flight which culminates in an out-of-the-body experience. If the energy remains blocked in the low-back or genital region, then the dreamer finds himself indulging in an ordinary sex dream of his own making. I find this a fascinating discovery, since yogis have for centuries used such disciplines as yoga and meditation to raise the "serpent power," or kundalini as it is called, from the base of the spine right up to the highest energy center, or chakra, situated at the top of the head, where it can be used for all kinds of higher spiritual activities.

DREAMS OF NUDITY

It is very widely believed that dreams of being naked or scantily clad in public are indications of sexual feelings or guilt about sex, but the truth is that in most cases such dreams have no reference to sex at all, and even those that do often have only an incidental concern with it. In the first place, such dreams can be literal warnings of something wrong with your clothes—and if you dream of finding yourself naked at the airport, do check that you have packed your pants for tomorrow's journey! If all is well at this level, then you must ask in what way you feel naked, revealed, vulnerable, exposed, or open in your life at the present time.

A university lecturer I know has a recurring dream in which he is walking through the college grounds or reading in the library, when he suddenly senses all eyes upon him. Looking down, he discovers to his dismay that he is naked or clad only in shoes and socks. As the dream takes place at college, it obviously refers to some aspect of his work, and he is able to relate it to the fact that he blatantly uses other people's ideas to advance himself, a habit he consciously considers rather clever. The dream, however, which usually occurs soon after he has published a paper, expresses his heart's fears that this time he is sure to be found out and "exposed" as a fraud—and the dream will no doubt recur until he gets head and heart together on this issue.

Sara, whose dream glossary appears in the Appendix, reported a similar kind of "exposure" dream, but for her this was an isolated occurrence and not a recurring theme, for the simple reason that she did not live in almost constant fear of exposure, as did the lecturer. After meeting us at a dream group in Pennsylvania, she asked us to let her know when we were coming north again so that she could arrange a private group session. We arrived in Baltimore unexpectedly, and Sara offered to arrange a small group at very short notice. That night she dreamed she was standing naked in a hospital room, when the door was flung open and a procession of people led by a guide filed through the room on a tour of the hospital. Taken by surprise, Sara tried to hide her nakedness by running to her bed, stopping to pick up a small piece of paper lying on the floor as she did so, and complaining bitterly to the guide of the unannounced invasion of privacy. The dream showed in beautiful picture language Sara's ambivalence about our unexpected arrival. While she was consciously delighted to have the chance to work with dreams, her heart was angry, for she would now be exposed as the fraud she felt herself to be—she had not been writing down her dreams as she had promised at our last meeting, and the most she could hope for now was to pick up one small dream (the piece of paper) before the weekend. The knowledge of her heart's thoughts put her in touch with her anger, which if left unrecognized might have sabotaged or spoiled the weekend for all of us.

Young men often ask me what it means to dream of being without trousers in public, and questioning usually reveals a concern with convincing the world of one's sexual prowess when underneath there lurks a fear of being exposed as sexually inadequate. A member of one of our dream study groups dreamed of seeing her teacher in the nude and being struck by his tiny penis. Her conscious mind admired him but the dream revealed her heart's feeling that underneath he was not much of a man.

Feeling tone is particularly important in getting to the meaning of nudity dreams. For example, one of our young college students dreamed that he disrobed in front of a cheering crowd of college friends. Here there was a reference to sex in that he had just experienced his first sexual intercourse with a girl, but even so this was only incidental to the dream's meaning, which was that he had managed to "shed" his moral prohibitions and felt delighted about it. Had the onlookers in the dream been disapproving, this would have indicated guilt feelings, for in the objective world, his fellow students would certainly have approved. On the same principle, a dream in which you find yourself embarrassingly exposed but no one takes the slightest notice, is a message from your heart that some disclosure you are consciously very concerned about is really nothing extraordinary.

Honesty, openness, and vulnerability are often symbolized in dreams by nudity. I recently fell asleep debating whether or not my small daughter was telling the truth about something and dreamed of her standing in front of me quite naked. The dream corroborated my conscious feeling that she had told the truth, and my heart was confirming my hunch about her honesty. By this, I am not implying that the heart is necessarily always correct: I am simply saying that in this case it had not picked up contrary vibes during the course of our encounter. As a general rule, however, I have found that when head and heart agree on any issue, there is a good chance of their being correct, whereas if they disagree you had better start asking questions.

EXAMINATION DREAMS

The commonest type of examination dream is one in which we sit down at a desk, look at the paper, and realize to our horror that we cannot answer a single question. Less frequent are the dreams in which we feel we have passed the examination successfully. The first question to ask on waking from an examination dream is whether or not there really is an impending examination or test confronting us in the near future, for if there is, then the dream is a clear warning to do more study or survey the situation more carefully. This can be very useful if we consciously feel quite confident of our chances.

The majority of examination dreams, however, are metaphorical, expressing our heart's thoughts that we are "under examination" or being "put to the test" on some issue in our present life, usually with the fear that we shall not make the grade. It is not at all surprising that such dreams recur time and again in the lives of most people in our society, for we are taught from the cradle to view life as "one big test" in everything from getting our parents' approval, to passing real tests and examinations throughout our schooldays and beyond, to making the grade in a competitive adult society, and living up to our own inner ideals. So it is not enough to say that a particular examination dream shows the dreamer feeling tested in his life generally: he would be very unusual if he did not. It is essential to relate the dream to specific circumstances, for this will enable him to see in what particular area of life he feels strained and hopefully point the way toward some constructive solution.

For example, I recently dreamed of passing a history examination and failing an English test on linguistics. As I was totally absorbed in writing this book at the time, I had no difficulty in relating the history examination to my review of the history of dream research in *Dream Power,* and the linguistics test to the next chapter, on puns, slang, and metaphor. My heart was warning me in no uncertain terms that while I had done a good job on the former, there was quite a bit to be desired on the latter, so I called a professor of English to check out my facts. I was exceedingly grateful to my dream, for I had been inexact on several points. So while it would be perfectly true to say that I feel myself under examination on this book—hardly surprising when one bears the reviewers and critics in mind—such a general interpretation would never have inspired me to "pass" the specific linguistics test, which was the purpose of the dream.

On the evening prior to giving a lecture on dreams last year, I dreamed that I walked into an examination hall to take a biology examination. It suddenly occurred to me that I hadn't done any biology for years and would probably fail the test. So, with an unusual spurt of dream confidence, I confronted the biology instructor and said, "I will look at the questions, but if I can't answer them I shall leave." The significance of this dream did not dawn on me until the start of my lecture when a woman biologist interrupted with what I considered irrelevant questions about the biology of the dreaming process. Realizing that my dream was a specific warning on this issue (we had already been in conference for two days, so my heart had time to pick up subliminal vibes from the other participants), I at once told the dream to the group, adding that I would answer any questions if I could but reserved the right to leave them if I could not. My dream saved much time and energy that day, and I felt really good about my newfound firmness and boldness in dealing with my "examiners."

To religious people, examination dreams often reflect their feelings of failing or "passing the test" in their spiritual growth, and this is probably what lies at the back of the widespread belief that dreams of being unprepared for an examination indicate the dreamer's unpreparedness for death and the "final judgment." I regard this whole way of thinking as thoroughly unhealthy, indicating that religion has been overtaken by the very disease it is meant to cure. True spiritual progress involves learning to sit lightly to the world's demands for competition and becoming as unconcerned with working for approval as the lilies of the field. On this basis, the best indicator of spiritual growth would be a reduction in the number of examination dreams we experience, not whether we pass or fail them.

DREAMS OF LOSING TEETH

If you dream of losing your teeth, check your mouth carefully first to discover whether or not your teeth really need attention. So often in its replay of the day's events in depth, the dreaming mind throws up subliminal perceptions of wobbly or decaying teeth, or of developing abscesses which we have been too preoccupied to notice or which may even be too subtle to be consciously registered by the waking mind. I take such dreams very seriously as literal warnings, for I have more than once been saved serious embarrassment by doing so.

If, however, your teeth are in good order, then you must ask yourself what feeling your dream is expressing, and this will depend on what teeth mean to you. My own loss-of-teeth dreams almost always reflect my feeling that I have "lost face" or "spoiled my self-image" in some way during the day, usually by giving in to emotions of fear or weakness. To call this "castration anxiety" (fear of losing one's masculinity, or in the case of a woman, her pseudo-masculinity), as the Freudians do, is beside the point, for many other people have loss-of-teeth dreams when they have spoiled their very feminine, passive, nurturing self-image.

To Edgar Cayce a dream of losing teeth meant loose or careless speech, a dream of false teeth signified falsehood, and a dream of infected teeth referred to foul language. But this cannot be taken as a universal rule, any more than the Freudian interpretation. To many people, teeth symbolize aggression, while to Stephen Dedalus, hero of James Joyce's novel *Ulysses,* they stand for decisiveness, and in the novel's dream sequences his loss of teeth symbolizes his loss of the power of decisive action. I have also come across cases where dreams of losing teeth symbolize "growing up" in the sense of maturing to a new stage of life, presumably based on the dreamer's memory of milk teeth falling out in childhood.

So if you dream of losing teeth without real dental trouble, then ask yourself what your teeth mean to you—potency, aggression, appearance (self-image), decisiveness, or whatever—and then try to discover what it is in your current life that is making you feel "toothless."

DREAMS OF LOSING MONEY AND VALUABLES

If you dream of losing your wallet, money, or other valuables, always check the next morning to make sure that you still have them. You may have lost something the previous day without consciously realizing the fact, in which case your dreaming mind is throwing up your subliminal perception in drama form to warn you to take action before it is too late. I dreamed one night of losing my wallet, and the following day discovered that it was in fact missing. I had been on the nearby beach the previous evening, so I immediately searched in that area and found it. Had I not recalled the dream, I should probably not have discovered the loss for several days, by which time it would have been too late.

Another down-to-earth possibility is that such dreams are warnings of likely future loss or theft. A friend of mine dreamed of losing her wallet, but found it safely inside her handbag the following morning. In looking for it, however, she noticed that her handbag was coming apart at the seams and realized that her dream was warning her to repair it before actual loss occurred. She had evidently noticed the tear in the bag without consciously registering the fact.

If, however, you have neither lost, nor are in any danger of losing money, valuables, or possessions, and you dream of doing so, then you should ask yourself what *values* you feel you may be losing in your life. A student who participated in my original research experiment on dream recall in Britain presented me with a dream of losing the rose from her ring. The ring was a treasured possession and she was greatly relieved on waking to find it still intact. Knowing something of Freud, she herself interpreted the dream as a fear of losing her virginity (deflowering), and related it to the fact that she was thinking of moving in with a boyfriend. On further discussion the following morning, however, it emerged that it was not so much the actual loss of physical virginity that worried her, as the sense that student life was making her lose the cherished values of her family upbringing.

When I appeared on *What's My Line?* in New York, an ex-Miss America told a recurring dream of running across a busy street, dropping her wallet, and dashing back to the curb to retrieve it. While we had no time to discuss the dream in depth, my guess was that she felt herself to be in the process of transition (crossing the street) from unknown girl to celebrity more quickly than she anticipated, and in her hurry to reach stardom felt herself to be in danger of dropping some of her values. These need not necessarily be sexual values: a sudden transition to prominence can threaten the domestic values of home life, moral values on account of the scheming that often seems necessary in a public career, or even spiritual values such as warmth and sensitivity. However, this dream has a happy ending, for it shows her running back to the curb and retrieving the wallet, which suggests that she feels able to "curb" herself in time to save the values in question.

It is important to remember that a feared loss of values may not necessarily be something to be avoided: often old values have to be lost so that we may grow by finding new and deeper ones. In a series of dreams that changed my life I dreamed successively of losing my wallet, handbag, money, books, clothes, and home, and while in every dream I was initially heartbroken about the loss, I found myself reflecting in the dream itself that it was not the end of the world, since I had ample resources to cope with life independently of these things. These dreams marked my slowly growing recognition —arrived at mainly from my work with dreams in my own personal life—that I am not dependent for my essential identity on playing any role, whether that of careful financial manager (as my mother had urged me to be), householder (my father's ideal), author, psychologist, or public figure. This discovery of the essential core of inner selfhood is obviously closely related to what the world's great spiritual disciplines describe as "detachment"—detachment from social roles in favor of a deeper reality which might be called the divine essence—but in my case, this initial realization came not from sitting at the feet of any guru, but from the "guru within" who speaks to us every night in dreams if we take the trouble to listen to him.

DREAMS OF FINDING MONEY AND VALUABLES

Finding money seems to be a major theme of the great American dream, for practically every time I took part in an American TV or radio show, at least one member of the studio staff asked me about a dream of finding money. As they were all invariably short of cash at the time of the dream, they feared it might be mere wish fulfillment, but I assured them that there was usually a lot more to it than that. For example, how much money did they find? Where did they find it? And what did they do with it?

One producer related a dream of finding a pile of gold coins beneath the foundations of a house he was thinking of buying. As there was not much likelihood of real buried treasure, I asked him about the house. "Well, it seems rather expensive at the moment," he said, "especially as we are short of cash and my wife says we can't possibly afford it. I've been trying to tell her that it will rise in value over the years. . . ." His wife and colleagues were inclined to dismiss the dream as wish fulfillment designed to prove his own point, but when we explained that his heart may possibly have picked up vibes about the place, this resonated immediately with his inner conviction that the house's situation would make it a "veritable gold mine" in years to come when waterfront property became scarce. His feeling may, of course, turn out to be mistaken, but this is true of any financial venture; the dream assured him, however, that his heart agreed with his head, which in my experience is sufficient reassurance for following through a particular hunch.

If there is no prospect of literal financial gain facing you, then a dream of finding money or valuables may reflect an inner feeling of your own "value" or "worth" as a person. A ten-year-old girl I know has a recurrent dream of being operated on, but instead of the surgeon's finding some disease, he always finds a precious stone in her body. We traced the last dream of this kind back to a family row in which her mother called her a monster and "rotten at the core," which we gathered was a fairly common accusation. It was obvious that her heart was protesting against this judgment, reflecting her very healthy feeling that she is intrinsically good, despite all outward appearances to the contrary.

The same applies if a dream of finding money or valuables occurs at a time when the dreamer's external life is going through a bad patch. Such a dream is a reassurance from the depths of the psyche that there are "inner riches" and resources in the personality which will insure that any impoverished condition will soon be overcome. In corroboration of this point of view I received this morning a letter from Australia in which a young woman described the anxiety and loneliness she felt on separating from her husband. These feelings were reflected in recurrent dreams of falling from a boat into deep water and waking in fright. She wrote that, in the final dream of the series, "I was on a boat again. This time I was not alone. Lots of people were falling overboard and while I saw them sink down deeply into the water, they did come up again. Then I fell, but instead of waking in fright, I landed safely and found myself happily picking up silver coins! I woke up feeling sure that the level of anxiety about loneliness, which I now recognize as pathological, was cured and that I will not experience it again, even though the outer circumstances of my life may not change."

In all the above examples the dreamers were delighted with the discovered treasure and accepted it gladly, but this is not everyone's attitude to a dream windfall. Often the dreamer feels he has no right to what he has found, or at least not to all of it, and he determines to hand it over to the authorities. Dreams of this kind have nothing to do with the ethics of real treasure discovery: they reflect a psychological problem of self-deprecation, a feeling that one's own worth is so low that any kind of good fortune must be rejected. Such dreams are important warnings of an unhealthy attitude which could lead us to sabotage our own successes, and I shall be dealing at length with this problem in Part III of the book.

SEX DREAMS

I am using the term "sex dreams" with its simple meaning here, to denote dreams depicting overt sexual activity or explicit sexual feelings. The fact that some apparently nonsexual dreams turn out to be symbolic representations of sex (poking a cow with a gun, or turning a key in a lock, to take just two commonly cited examples) is a different question altogether, and shows you how you feel about sex in your life (in the above cases, as an aggressive act toward someone you see as a "cow," or as the unlocking of new possibilities and opening up of a relationship).

Popular thinking about overt sex dreams has suffered from the Freudian belief that dreams are wish fulfillments—witness the myth that virile men are always dreaming of having Raquel Welch on a tropical beach. The truth is that we do not dream randomly about sex any more than about other subjects. In fact, sex is like any other dream theme—it has a literal meaning if it reveals something about your actual sex feelings toward real people in your life at the time of the dream, but otherwise has to be understood as a metaphor for being "excited," "worked up," "turned on," "intimately involved," "frustrated," "deflated," or "intruded upon," and may refer to a cause you are "embracing," an idea you are "getting close to," or the "coming together" of two aspects of your personality. In these cases the metaphor does not tell you how you see sex, but how much libido you have invested in something in your life.

If your dream shows you being sexually involved with someone in your present life toward whom you have no conscious feelings of attraction, then it is almost certainly a straight warning dream. This is an area where society has benefited enormously from the Freudian revolution, which has made it possible for us to recognize that everyone has sexual feelings about other people all the time, including those of the same sex, children, and blood relatives. Such feelings and the fantasies that go with them are a normal part of life and do not in the least imply that the person who has them is sick or is a lascivious monster whose "real" desire is to break out into promiscuous or incestuous behavior. It is when we thrust such feelings and fantasies right out of conscious awareness that they become dangerous, by building up tension which can drive us to do things we have no true wish to do at all—and sex dreams often come to alert us to this danger.

A woman asked the meaning of her recurrent dreams of having sex with her next-door neighbor, which disturbed her because she was happily married. Having ascertained that the neighbor was reasonably attractive, I told her the dream meant simply that he turned her on sexually, whereupon she exclaimed in horror, "But I love my husband!" She seemed very relieved to learn that there need be no contradiction between loving a partner and enjoying sexual fantasies about others. I advised her to view the whole thing with a sense of humor, enjoy the dreams, and if possible tell her husband, who might be similarly worried by his own sexual fantasies. The most important thing in cases like this, however, is to become aware of one's own feelings, thereby avoiding the danger of the repressed impulses seeking expression in some devious way, perhaps by outbursts of anger against one's partner or by "accidentally" finding oneself in some compromising situation with the dream lover in waking life.

If you find yourself enjoying sex in a dream in circumstances your waking mind finds shocking, this is a sure sign that you are imposing a life-style on yourself which is at variance with your natural feelings in some way, and your dream is a warning to change it. In many cases the change needed is nothing more than a better sense of proportion about sex which will make you less uptight, and this can often be achieved by becoming aware of the fact that many of the ideas we accept as gospel truth are no more than outmoded and distorted opinions derived from parents, teachers, clergymen, and authority figures of the past. It is perfectly normal to have sexual feelings toward almost anybody, including members of our own family, and the mother who becomes aware of sexual feelings toward her child is in far less danger of becoming a Mrs. Portnoy than one who would "never dream of having such nasty thoughts." Similarly, the girl who is aware of being attracted to her father or brother runs far less risk of expecting boyfriends to be like him than if she were unconscious of her feelings. Many women in our society are shocked to dream of themselves as prostitutes until they understand that this is the heart's view of the way they have been trained from birth to give or withhold sexual favors in return for goods received. This very salutary discovery can bring about a much healthier life-style.

If your dream shows you sexually involved with someone who is not part of your present life, it must be using sex as a metaphor, and this is equally true whether it be a Raquel Welch-type dream or an incest dream of a parent or sibling no longer in direct contact with you. The first point to note in such dreams is what the sexual experience felt like, since this will help you identify the event of the day to which the dream refers. For example, a woman in one of our groups had a vivid dream of her long-dead mother raping her painfully with an enormous, bonelike penis, and said she felt furious at the sense of violent intrusion. Instead of trying to interpret this in terms of some transsexual Oedipus complex or similar psychoanalytic notion, we asked her whether she had felt the presence of her mother intruding on her during the course of the day. She resonated at once to this, recalling that she had been playing happily and noisily with her children when she had suddenly felt compelled to stop the game, calm them down, and tidy up the room. Her heart interpreted this experience as being "raped," "taken by force," and "penetrated" by her mother's strict views on child upbringing, and she realized how necessary it was to free herself from this piece of outmoded conditioning. This is a typical case of what Fritz Perls would call being ridden by a "topdog," and I shall be showing how to deal with such problems in Part III.

On the other hand, when I dreamed of making passionate love to Malcolm Muggeridge, the experience was unquestionably a pleasant one, but as he was not in my life at the time, the dream had to have a metaphorical meaning. (When we told him the dream later, his comment was, "But, of course —at my age it couldn't be otherwise!") Since he stands in my mind for very traditional Establishment attitudes and is also a frequent public performer, I was readily able to relate the dream to the good feeling I had the day before when I was well received (to my surprise) in a symposium where most of the other participants were rather traditional and uptight in their views. My heart was telling me that I was more "excited" than I realized by this "coming together," as I had feared they would reject some of my more way-out ideas. Similarly, when Sara (of the Appendix) dreamed of being in bed with a young black civil rights leader whom she had met only once in her life long ago, and of being frustrated because they never got to intercourse, she was able to relate this to her growing feeling of "dissatisfaction" with her job in a civil rights organization, which was concerned with housing low-income families.

Sara provided another example of a metaphorical sex dream which is one of the best in my collection. In the dream she was lying on the floor with a stranger whose name she knew to be

Hal, kissing and embracing, and very pleasantly engaged in foreplay. To her horror, she noticed that the walls were made of glass and hundreds of faces were watching them. She was too excited to stop but became frustrated because Hal seemed to be satisfied with merely rubbing his penis between her legs. She tried to help him, but he did not respond. In the end, he got up and walked off down the hall seeming very satisfied. Sara was left bewildered and frustrated by the whole experience, yet inexplicably felt she ought to thank him for the good time!

Sara called me because she could make nothing of this dream. Her only association was that Hal was the name of an impotent church friend, but he was not the man in the dream and she had not seen him for months. When I asked what had happened the previous day, she said that the legal suit she had brought against her organization for nonpayment of expenses had been settled very satisfactorily out of court, and everyone —her lawyer, parents, and church friends—was delighted because this meant that Sara was absolved and the organization admitted its responsibility. When I asked Sara for her own feelings, she said that she was happy too, though it had come as something of an anticlimax. "Like making love with Hal in a public place?" I asked, and we both laughed as the meaning of the dream became clear. Sara had, in fact, been getting very "excited" and "worked up" by the coming case, as she saw it as an opportunity to get her own back on the organization she felt had so mistreated her. She had a great deal of anger to come out, and although she was glad to get the money and be vindicated, she felt somehow "deflated," "let down," "unsatisfied," and deprived of her "orgasm." She was then able to identify Hal as a symbol of her lawyer (who also belonged to the church), revealing her heart's feeling that he had acted like an impotent man in allowing his love of peace to settle out of court and avoid an "un-Christian" conflict, when her need was not just to get the money but to discharge all her pent-up anger. Of course, she had thanked the lawyer after the settlement, but her heart had been fuming with frustration at the "anticlimax" of the whole "affair."

On the night after Sara and I had worked this out I had a vivid short dream of delightful sex with a young student who had been one of my subjects many years ago when I was doing experimental dream research with the EEG machine. A random sex dream? Not at all. It showed that I had been very excited and turned on by the resolution of Sara's dream—or, in the dream's language, I had felt very "sexy" about "my subject," namely dream interpretation. I then recalled how John's business colleagues would often talk of getting a sexual kick out of some new deal or scheme, how journalists often referred to feeling sexy about new ideas for an article, and how Janis Joplin said that when she performed with a rock band "it was better than it had been with any man!"

A strict Freudian would probably say that this shows that all our creative activities are sublimated sex, but I find it more meaningful and less down-putting to say that all our energies have an erotic character; the drive toward reproductive sex is just one of many possible manifestations of the basic life-urge to "pour ourselves into," or "embrace," or "take into ourselves" whatever excites us throughout the fibers of our being. Our ordinary waking consciousness represses most of these erotic feelings, but it is well known that under the influence of psychedelic drugs, when the brain's information processing is greatly speeded up, people can become conscious of relating to the entire environment in this way, with an experience of superorgasm whenever life-energy is able to flow freely. I believe this is why mystics have so often used erotic language to describe their experiences of total oneness with reality. It is possible that dreams are able to reveal this aspect of experience to us because they also involve the speeding up of the brain's information processing, which would explain the fact that the sex organs become excited during REM periods irrespective of dream content. And when our dreams use sex as a metaphor, they tell us just how much libido we have tied up in any particular event in our lives and what experiences set the life-energy flowing for us.

IDENTIFYING YOUR DREAM THEMES

If you have been keeping a dream diary as I suggested, then it would be useful at this stage of the dream game to examine the general themes of the dreams you have collected. Each page should be dated, the dream numbered and written out in full, followed by a brief description of the day's events and associations. The next step is to identify—in one sentence—the theme of each dream, as this often gives the clue to the dream's meaning by suggesting what further questions to ask. Here are a few examples to illustrate what I mean:

1. *Theme:* I am exploring an old house and find lots of unexpected rooms.
   (Question: What inner or outer possibilities am I exploring at the moment which may be larger than I thought?)
2. *Theme:* A tidal wave threatens to overwhelm me.
   (Question: By whom or what do I feel overwhelmed at the moment?)
3. *Theme:* A burglar climbs in through my window.
   (Question: Who or what am I trying to keep out of my life at the moment?)
4. *Theme:* War is declared and fighting breaks out.
   (Question: In what way do I feel my present life to be a battleground?)

In all the above examples the dreaming mind is working very much like a newspaper cartoonist, picking up common figures of speech and making a literal picture of them. Dreams often carry this process to extraordinary lengths, indulging in the most outrageous puns, which leave people incredulous when they first encounter them, and this is such an important part of dream language that it is worth a chapter to itself.

«6»

PUNNY THINGS IN DREAMS

A pun is a pistol let off at the ear; not a feather to tickle the intellect.
—Charles Lamb

On the night before my book *Dream Power* was published in New York, I dreamed of a man in long white underpants shooting me down with a machine gun. As I had to appear on a breakfast show at six-thirty the following morning, I had no time to ponder the meaning of my dream. In fact, I totally forgot about it amid the flurry of TV, radio, and press interviews that followed, until I found myself later that evening in the studio of Long John Nebel, who had invited me to be his guest on his four-hour live radio show. As I waited for the show to start, I suddenly realized that the man in my dream had been dressed in "long johns" and symbolized none other than Long John Nebel himself. I panicked as I now recalled rumors of his critical attacks on guests which had caused many of them to walk out halfway through. My conscious mind had forgotten all this in the excitement of publication events, but my heart remembered every word and threw up a warning in my dream to beware of being "shot down" the following day. As I had never met Mr. Nebel previously, my dreaming mind pictured a figure in long johns to make its point.

It was too late to cancel my appearance—and in any case, my dream merely reflected a *fear* that he would shoot me down—so I gritted my teeth and awaited the worst, determined to stick it out to the end, come what may. What happened more than justified the warning, though it was not Mr. Nebel who did the shooting. As he himself did not feel qualified to discuss dreams in depth, he had brought in a psychoanalyst friend as cohost, and I looked forward to a rational, scientific, and stimulating interchange of ideas. To my astonishment, Dr. S. immediately embarked on a virulent attack on my daring to criticize the great master, Freud, whose ideas (according to him) were now firmly established. When I asked for evidence of this, he quoted an out-of-date paper written by a Freudian analyst in the early 1960s, of no scientific validity whatsoever, and when I corroborated my arguments with genuine scientific authority, he called me an ignorant Britisher and penis-envying woman! The whole evening continued in this vein, with his opposing all my ideas not with the scientific honesty I had hoped for, but with vicious personal attacks of no relevance whatsoever to the subject of dreams. I left the show at midnight, a sadder but wiser person, little realizing that this was only the first of several such interviews with psychoanalysts of various schools who became quite irrational when confronted with scientific evidence that threatened their pet theories.

That night I dreamed I had been invited to take part in a game of cricket, but when I arrived, found myself involved in a very rough game of baseball, presided over by Long John Nebel and Dr. S. dressed in elaborate gilt uniforms. The dream expressed in beautiful picture language how I felt about the evening's events: I had been invited to participate in what I believed would be fair, open, rational discussion of dreams, but my hosts were "guilty" of tricking me into a "base" game of underhand moves, snide remarks, and devious attacks, which to me was definitely "not cricket."

People are apt to be incredulous when they first discover the dreaming mind's capacity for playing these kinds of tricks with words, and this is probably the aspect of my own work that has most attracted popular attention, so much so that such dreams are nowadays often called "Faraday dreams," although it was in no sense an original discovery on my part. Freud himself observed it in his work in Vienna at the beginning of the century, and it has since been noted by Jung and other dream experts throughout the world, including Edgar Cayce in America, which proves that it is not just a byproduct of the British sense of humor, as many Americans suggested.

The truth is that while different nations may make more or less use of punning in professional humor, the device itself—that of associating disparate ideas because the words for them happen to sound the same—is one of the most basic elements in all human thinking. Arthur Koestler has even gone so far as to argue in his book *Insight and Outlook* that jokes represent the primitive basis of creative thought, because they bring together ideas that are normally considered unrelated—the essential factor in all real originality. Maybe the British sense of humor is more primitive than the American, but I can now testify from experience that Americans are just as prone to making puns in the stream of thought that goes on at the back of their minds and comes out in dream pictures.

While it is not possible to make hard-and-fast distinctions in this area, I think it is worthwhile identifying at least six different kinds of punning in which the dreaming mind is wont to indulge:

1. Dreams based on *verbal* puns in which one word represents another of similar pronunciation but different spelling—for example, my dream of men being dressed in *gilt* to express a feeling of their being steeped in *guilt.*

2. Dreams based on *reversal* puns—for example, a dream of

filling full a jar which expresses a sense of being *fulfilled.*

3. Dreams based on *visual* puns in which the dream creates a picture based on one sense of a word in order to express an idea involving a different sense of the same word—for example, my dream of a *base*ball game to reflect my feeling of being involved in a *base,* underhand game.

4. Dreams based on puns involving *proper names*—as when I dreamed of a man in long johns to represent Long John Nebel.

5. Dreams which create a literal picture of some colloquial or slang metaphor—for example, when my dream depicted a man "shooting me down" to express my fear of being attacked verbally, and a cancelled cricket match to express my feeling that something was "not cricket."

6. Dreams which create a literal picture of common body language—for example, a dream of a bare chest to depict a feeling of "getting something off one's chest," and a dream of a one-armed man to reflect a feeling of being "disarmed."

In all cases, the overriding rules of interpretation are those given in the preceding chapter, namely that the dream's meaning lies in its reference to an event or thought of the previous day or two, it comes to tell us something we do not already know, and it is correctly interpreted when it makes sense to the dreamer by motivating him to change his life in some constructive fashion.

DREAMS BASED ON VERBAL PUNS

While spending a weekend in a Long Island commune I made the acquaintance of one member whose rude, aggressive behavior was so disturbing to the other members that they were seriously thinking of asking him to quit. That night I dreamed he went around the commune slashing us all with razor blades, and as I tried desperately to escape in the dream, I noticed that his eyes were very red and puffy, as though he had been crying. It suddenly occurred to me in the dream itself that he had a sore eye, which I immediately associated with "sore I"—that his aggressiveness was a byproduct of some deep hurt done to his ego by someone or something in his life. When I shared my dream with some of the commune members the next morning, they resonated with the diagnosis and decided to talk things over with him instead of just asking him to leave the commune.

I have since used the verbal pun of "an eye for an I" many times in my dreams. Whenever I see an animal with golden eyes, I know my dreams are reminding me of the essential goodness, the pure gold, at the center of our much-maligned "animal" nature. Closed eyes reflect my feeling that I or someone else in my life has a closed, uptight attitude to something, a closed "I" unable to open itself to new possibilities. And when I dreamed of a nun with shining red fiery eyes, I realized that someone in my life at the time was concealing a very warm, passionate "I" at the center of her personality beneath a cold, nunlike exterior.

I recently dreamed of entering a restaurant and asking to see the menu. There was only one dish being offered—wild boar—and the meal, which included one free drink, cost $12. I thought this was exorbitant and decided against it. I did not understand the dream until I settled down to my day's work, which was to write a book review for a journal. As I typed out the title and publisher's name, I noticed that the price of the book was $12, which immediately reminded me of my dream. I had been reading the book before going to sleep and wondering what I was going to say about it. My heart presented its feelings on the subject very clearly—that while the "food for thought" contained therein had a certain amount of "spirit," the heavy presentation was a *"bore"* and certainly not worth $12. As this resonated with my conscious feelings about the book, I had no difficulty in quickly completing the review.

Other nice examples of verbal puns appearing in my records are "reversing the Rolls" for "reversing roles"; and a dog "gnawing at the sole" of a sandal to reflect the feeling of something "gnawing at the soul." Perhaps the most ingenious was supplied by a British executive who dreamed that he was wearing a "surplice." When we asked if he was intending to enter the church, he replied, "No, I guess it reflects my fear of being made redundant" ("surplus").

DREAMS BASED ON REVERSAL PUNS

At a dream group in Baltimore I was given an example of a different kind of verbal pun, a picture based on syllable reversal. Donna, a student, dreamed of driving down the fast lane of a highway at 70 mph and refusing to move to the slow lane so that the honking car behind her could pass. She was unable to resonate to the group's suggestion that someone in her life was trying to "overtake" or pass her by, but she responded immediately when her tutor asked whether or not she was standing out against someone she felt was trying to "take her over." The culprit turned out to be her father, who was interfering in Donna's life by pressing her to live up to his expectations of what a "good girl" should be, and she was vehemently resisting his efforts to convert her into a "little lady." "Stand up, Donna, and stop slouching," he would say. "Don't sit with your legs crossed, and wear skirts instead of those old jeans . . ." and so on. Donna felt this to be a violation of her essential self, and the dream showed that both her head and heart were together in their firm refusal to allow her to be "taken over" by her father's powerful, "driving" personality.

Roy dreamed that a bulldozer was pushing its way through the rooms of his house. As there were no bulldozers, either literally or figuratively, in Roy's life at the time, we all agreed that this must represent a part of his own personality, a conclusion we found puzzling as Roy is in no way an overtly aggressive person. As we discussed the dream with Roy and some friends at their Thanksgiving party, someone hit the nail on the head by exclaiming that Roy was more like a "dozing bull" than a bulldozer. A howl of resonance filled the room; everyone, including Roy himself, agreed that the image of a dozing bull fitted him exactly. He gives the impression of a kind of sleepy Ferdinand who prefers to smell flowers rather than quarrel, who at the same time sits on a vast store of smoldering energy which could one day erupt and cause him to charge. My hunch is that Roy is well aware of this energy and fears it, preferring to use his "dozing bull" image to bulldoze his way through life, in order to get his own way without causing overt harm to others. He is now known affectionately by his family and friends as "big chief Dozing Bull."

Other examples of verbal puns based on syllable reversal in my dream collection are "standing under" to signify an "understanding"; "looking over" warning of an oversight; and the image of an ironing board to indicate that the dreamer was feeling "bored stiff" with a present project in his life.

DREAMS BASED ON VISUAL PUNS

The two preceding sections have shown the dreaming mind creating pictures based on word twisting. Dreams also create purely visual puns, however, in that they produce pictures based on one meaning of a word to express feelings or ideas involving the same word in a different usage.

At the Long Island commune mentioned earlier in this chapter, four of us agreed to try to dream about each other and share our dreams the following day, as we felt it was essential for people living under the same roof to know how they really feel about each other. I was very embarrassed to dream of Stan —a handsome, charming, kind, and intelligent man—as a German Nazi, which implied that my heart had picked up vibes of possible ruthlessness and cruelty. He, in turn, dreamed of the three of us as characters in the play he was writing and appeared to gain quite a bit of insight from the dream. We forgot about the dreams until a week later, back in New York, when Stan's girlfriend called us in tears, saying, "Ann, your dream was right. The man's a monster. He came with me that weekend only to get revenge on another woman and make her jealous." While I comforted my friend and congratulated myself on my powers, not of extrasensory perception but of subliminal perception, I suddenly saw that Stan's dream had a similar message for us. It was a punning statement of the fact that he was just "playing" with us all, in the sense of getting us to take part in a drama he had set up for his own purposes.

Many visual puns appear in dreams as recurring symbols to express a particular idea or feeling. For example, whenever an author friend dreams of a roll of material, he knows the dream is saying something about "material" for his latest book. A roll of old cloth expresses his feeling that something is "old stuff"; new cloth reflects his feeling of having found "new material" for the book; and rough cloth means that he feels the material in question is a bit too racy for publication.

Similarly, whenever I dream of alcoholic beverages I know that my dream is saying something about "spirit." The absence of alcohol reflects my feeling that something in my life is lacking in spirit, and I have recently come to the realization that my recurring dreams of being a barmaid reflect my feeling of "serving the Spirit" in a religious context—a subject I shall return to later in the book. So whenever you find yourself on a train in a dream, do consider the possibility that you might feel yourself to be "in training" for something in your life, and, similarly, if you are arrested in a dream, bear in mind your feeling of possible "arrested" development in your life.

DREAMS BASED ON PUNS INVOLVING PROPER NAMES

Like many TV and radio personalities we met on our tour of the United States, Bill Gordon rarely recalls his dreams. However, when we appeared on his live TV show in San Diego, he triumphantly asked us in front of the cameras about a dream of the previous night in which he was dressed in fencing clothes with a mask over his face and was savagely slashing at the branches of a beautiful, delicate tree. We hazarded the guess that he must have been angry with someone the previous day, perhaps without realizing it consciously, and had "masked" his true feelings out of fear or expediency. He responded that he had indeed had an infuriating interview with one of his chiefs, and we considered the dream interpreted, except for one point. Why had the dream chosen to depict the chief as a tree? Was his name "Ash" or "Wood," or even "Branch" or "Leaf," or any other similar word to justify the dream's use of a tree as a symbol for the object of his wrath? As Bill could find no connection, we left the subject there.

On the way home, however, his producer asked whether we had heard a gasp from the audience when we mentioned the word "tree." This came from Bill's friend Theresa (pronounced Ter*ee*sa, or Treesa for short), who clearly recognized herself as the target of Bill's anger and related it to the fact that she had not been waiting for him after his traumatic interview with his chief. She knew he was disappointed and upset by her absence but had not realized the full extent of his anger. Nor had Bill. He had been so preoccupied with his interview that all other feelings had been banished, to emerge later in the form of a dream. The result was a beautiful name pun based entirely on sound.

Why did the dream depict Bill slashing at a tree rather than at Theresa herself? I do not believe that the dream was bent on disguising the painful truth, as Freud would have said, but rather that it used the symbol of the tree to express an agglomeration of ideas, including Bill's deeper feelings about Theresa. In his mind's eye Bill probably had a picture of Theresa as a beautiful delicate source of growth in his life, but totally defenseless and vulnerable to any attack he might make on her. The last word has to come from Bill himself—but all those who know Theresa agree that his dream image of her is a perfect one.

Sara, whose dream of nudity was described in the last chapter, gave us a very nice example of a name pun when we first met her at Kirkridge, a Christian retreat run by Jack and Jane Nelson. Sara dreamed of a woman called Grace Nelson, and she interpreted this name change as her heart's thought that Jane was a very "gracious" and "graceful" person.

The names we give our dream characters are often amusing, but they always make the point of drawing to our attention some particular quality of the person in question. For example, John dreamed of meeting a ghost called Ed Stynge in a church, and this dream fragment condensed into one single image his heart's thoughts about a certain friend of his—namely that he was "stingy," not in any monetary sense but in the giving forth of life-energy, as the disembodied image of a ghost implied. And when a friend of mine dreamed of a fictitious character called Mrs. Bagley she was able to resonate to the thought that the psychotherapist with whom she had talked the previous day was an "old windbag."

In all the above examples the dreaming mind created a picture or an imaginary character with a special dream name to reflect feelings about actual people in the dreamer's life, but there are occasions when the pun works the other way around. For example, David Frost often appears in dreams to express a feeling of coldness in some aspect of one's life, and I myself had a dream featuring Bob Hope and Dorothy Lamour which represented the concepts of "hope" and "love." This was capped in a recent lecture at the Association for Research and Enlightenment (A.R.E.), the Edgar Cayce foundation in Virginia Beach, by Elsie Sechrist (author of *Dreams—Your Magic Mirror*), who reported a student's dream in which Bob Hope came jumping down the street on a pogo stick, expressing the dreamer's feeling that "hope springs eternal"!

People who think of themselves and their friends in astrological terms will often produce dream puns based on astrological signs, even though they have no belief in the truth of astrology. If a bull appears in my dreams, it almost certainly represents John, who is a Taurus—a useful pun for my heart to employ whenever I have found him being stubborn or blindly aggressive during the day. The fact that I am a Gemini provides me with an obvious symbol whenever I am feeling split. So if you dream of a raging lion, ask yourself if you feel threatened by anyone in your life who is a Leo before you start speculating about its being an archetypal symbol of your own aggressive impulses, and do not interpret a dream of a virgin as a Jungian anima symbol before you have considered whether someone you know is a Virgo.

"DREAM CARTOONS" OF COMMON SLANG EXPRESSIONS AND COLLOQUIAL METAPHORS

I hope we shall never lose sight of the fact that dreams are often extremely funny—much cleverer and wittier expressions than we normally produce in waking life. Such dreams have the bonus of entertaining as well as instructing, and sometimes they can keep us laughing for days, which in itself is an immensely relaxing and healing process. I have noticed that such dreams often come at times when a little humor in our lives is needed. For example, I have just received a letter from a friend who dreamed on the night before a medical checkup for a possibly serious complaint that a group of twenty people were about to die, though none of them knew the others were in the same position. "Each one was going to die differently," wrote Jane. "Then I saw the girlfriend of my former lover of many years ago and when I asked her what *he* was dying of, she said, 'his legs,' he was getting crippled. This cracks me up because I immediately thought, 'of course, his legs—from all that *running around!*' At this point the nightmarish quality of the dream vanished and I felt like laughing!" This dream fulfilled two functions—to bring a sense of proportion to the situation by reminding Jane that we all have to die in our own way one day, and more important, to put her in touch with one of the greatest gifts of all, our sense of humor, which she badly needed at the time.

This particular part of Jane's dream can be likened to the humor of a cartoonist, and "dream cartoons" based on colloquial and slang expressions are very common indeed once we develop the art of spotting them. Reading through our students' end-of-term papers on the interpretation of their own dreams, I came across the first line of Dave's dream in which he wrote that "some friends and I were going to a camp in the woods," a perfectly innocent phrase that would catch the attention of only the most suspicious mind. Mine is such a mind when it comes to deciphering the messages of dreams, and it immediately turned over the possibility that Dave's dream was concerned with "camp," the slang word that can be used to describe homosexual behavior, though I was quite prepared to abandon the notion if it failed to resonate with Dave. However, I did not even need to ask, as the dream continued, "I gave two guys a ride. They turned out to be homosexuals. I tried to flee—one of them caught up with me. . . ." Dave concluded, "I view these dreams as expressing the fears I feel about homosexuality. I have never had a homosexual experience but I've often wondered without society's brainwashing whether I would flee from such encounters as I attempt to do in the dream." In this case, the slang expression of "going to a camp" was redundant, as there was overt homosexuality later in the dream.

The same thing happened in Al's dream in which his friend X came into the room chasing a ball and tried to get Al to join in. Al himself recognized the slang expression of "chasing a ball" as a picture of homosexuality and was able to relate it to the fact that X, who was bisexual, was trying to persuade Al to "join the game." Both Dave's and Al's dreams were "overdetermined" in the sense that the theme was repeated in different ways, not in an attempt to disguise the truth as Freud believed, but because the thoughts of the heart were doubly concerned to get the message across to the dreamers.

Lisa Hammel of *The New York Times* related what she called a "Faraday dream" in which she entered a room, sat down, and was horrified to see herself sitting in the next chair. She woke up realizing that she was "beside herself" about what she had hitherto considered a minor problem in her life. The dream showed just how upset she really felt about this and re-created the sense of being split into two which was the origin of the phrase.

Caroline, an assistant in my publisher's office, dreamed that she was standing in front of a woman who was trying to drag her sweater off over her head. Caroline woke up realizing "what I'd suspected for weeks—that this woman was trying to 'pull the wool over my eyes.' " Here again, the dream goes back to the origin of the expression, the use of woolen blindfolds for people being led to torture or execution. In another dream Caroline found herself about to take an enormous mouthful of Chinese food. As it entered her mouth, it burned her tongue so badly that she spluttered it out all over the table. She immediately related the dream to the feeling of having "bitten off more than she could chew" in taking on a new course of Oriental studies.

Mary, who together with Sara and Jane formed a small dream group in Virginia, dreamed of being a maid wiping the crumbs off the kitchen table while the family sat around eating breakfast and taking no notice of her at all. She woke up realizing just how "crummy" she felt about the housekeeper role she was assuming in order to please the family. I was delighted with the dream pun as soon as I heard it, but it was only after consulting a work of reference that I discovered the origin of the word: "crummy" derives from the early English "crumme," which is the same as "crumb," and acquired its common slang meaning by indicating a feeling that one is crumbling away like a piece of moldy bread or is covered all over with crumbs just like a messy table.

On another occasion, Mary dreamed that a friend whom she had just seen was covered with bugs and was astonished to learn that the friend had actually been taken to a mental institution ("bughouse") the night after the dream. Mary believed she had picked up subliminal vibes that her friend was not as well as she outwardly seemed to be. When she told me this dream, I recalled that bugs often appear in dreams to reflect the dreamer's feeling of being "bugged" by something or someone in his life. Here again, it was only when Mary's dream drove me to reference books that I discovered the word "bug" comes from the early English word "bugge," meaning hobgoblin or demon, and was later applied to insects, which can plague us like demons. So the dreaming mind is being entirely accurate in pictures like these, using insects to reflect anything from a feeling of annoyance or irritation to a sense of real mental illness, which was at one time attributed to possession by demons.

The recognition of puns often enables us to interpret dreams simply and immediately in a general sense, though further reflection is sometimes needed to discover the current specific situation to which the dream refers. An American comedian who has recurring dreams of driving fast and dangerously through heavy traffic has no difficulty in understanding the message to stop "driving himself" so hard and to slow down before he causes an accident. But despite the added warning that his life lacks direction (driving "in the dark"), he persists in his precarious life-style—and his dream will recur until he changes it.

A businessman dreamed that he was in a large room swinging a weight around at the end of a long piece of rope, making his colleagues jump over it. Until this time he had kept quiet about a certain work situation in order to avoid a confrontation, but now he took the dream's advice to "throw his weight around" more than he had been doing and to "make these people jump."

John, my husband, dreamed of watching Fred Astaire dancing with Rita Hayworth, when Frank Sinatra suddenly stabbed Fred in the back. John immediately resonated to the notion of jealous feelings but found it difficult to accept that such feelings in himself were becoming murderous. When I asked what the characters in the dream meant to him, he replied that they made a good song and dance act. We both laughed at the dream's message—that John was making a big "song and dance" about the whole issue and had been blowing his feelings up into a great big drama.

Other dream cartoons in my collection are "climbing up the wall," "left holding the baby," "seeing the writing on the wall," "being up the pole," "that's not my bag (or cup of tea)," "leading up the garden path," "getting to the bottom of something," "sitting on the fence," "changing one's tune," "having a finger in every pie," "keeping it under one's hat," "climbing to the top of the tree," "taking the bull by the horns," "my goose is cooked," "going down the drain," "skating on thin ice," "being tied up," "being washed up," "pulling one's socks up," "nailing someone down," "being too big for one's boots," and many more, some of which will appear later in the book. In waking life we toss off such expressions without (usually) a thought of their literal meaning, but the heart registers them precisely because the pictures they invoke bring home to us in the clearest possible way the full intensity of some current life situation.

DREAM PUNS BASED ON BODY LANGUAGE

Among the vast number of metaphorical figures of speech that make up our language probably the most fundamental are those which express feelings in terms of bodily attitudes, like "keeping one's chin up" or being "tight-fisted," for these often describe actual somatic changes that accompany the feelings in question, usually without our being consciously aware of the fact. Over the past decade it has become standard practice in most forms of psychotherapy to take this interrelationship between mind and body very seriously, by looking at what people are "saying" with their bodies as well as listening to what they say in words. I believe we should also take special care to spot dream puns based on body metaphors, for they are likely to be drawing attention to feelings that will influence not only our mental well-being but our physical health as well, by bringing about psychosomatic effects.

My first husband, who is an osteopath, noted years ago how "heartbroken" people are prone to develop heart disease, how respiratory ailments are often caused by something the patient cannot "get off his chest," how neck trouble can frequently be traced back to someone in the person's life who is a "pain in the neck," and so on. In one beautiful case a patient who had developed a squint at age five when his mother ran off with another man was totally cured when he talked to my husband and eventually "*faced* the truth" that she would never return. Medical science in recent years has begun to uncover some of the mechanisms by which this kind of psychosomatic influence takes effect, and it is now considered a much more important factor in physical health than was ever contemplated by the mechanistically oriented medicine of the last century.

A nice example of a dream pun based on body language was given us recently when we had dinner with friends who wanted us to meet their son, Jay, and daughter-in-law, Cindy, who had been married for three years. Jay and his parents are very much of the new order, holding way-out views on sex roles in society, the family, and life generally, which is difficult for Cindy, who was raised in the old traditions by a Southern mother. As we talked, it was obvious that she was trying really hard to be "with it" but felt very threatened by the whole situation. She related a recurring dream in which she has a baby who can talk but who spits out the pieces of meat she feeds it. She believed the dream had something to do with her desire for a baby and joked about how all mothers expect to produce child prodigies. While this wish may have served as a trigger for the dream, it was abundantly clear to all present that the dream presented a beautiful picture of Cindy herself, who despite her brave *talk* of the new culture feels herself very much a baby in a world she does not understand and who is unable to assimilate ("swallow" and "digest") these tough ideas. The dream was Cindy's way of articulating feelings of which she was not fully aware and a plea to the family not to push their trip on her too soon. I am strongly inclined to believe that if a new life-style is forced on her before she can "stomach" it, she will be in danger of falling physically sick. (As a postscript to this story, Cindy told us at our last meeting that she was "fed up" and "sick and tired" of being treated like the "baby" of the family and was insisting that her own needs for adulthood, which included having a baby, be met.)

Sometimes the dreaming mind goes as far as depicting actual illness to make its point. I was rather worried by a call I received on a radio show in New York from a woman who said she had a recurrent dream of vomiting up a lump from her stomach. As in all cases like this, I advised a physical checkup, for the dreaming mind might be detecting a disorder before it became apparent in waking life. However, the caller seemed confident that there was nothing wrong with her physically but was able to relate the dream to the fact that she could not "stomach" the fact that her husband had a mistress. In fact, the dreams began soon after he had told her about it. I advised her to talk things over with him honestly and state her feelings, before she became physically "sick." I have indeed known people to die when they became "sick to death" of something or someone in their lives.

In a similar way, a member of one of our British workshops dreamed of having an operation in which his heart was found to be defective and was replaced by a beautiful, tiny plastic heart. The surgeon told him that it would feel a little cold at first, but he would get used to it. In this case there was no need for any physical checkup because he had had one the previous day; this probably served as the trigger for the dream, but certainly did not explain it, since his heart and whole body had been pronounced in top condition. When we asked if he felt himself being "cold-hearted" in any way, he immediately resonated to the feeling that since he was contemplating a divorce, he must have a defective heart. By bringing the problem fully into conscious awareness and recognizing rationally that breaking up an unsatisfactory marriage was not a "cold-hearted" act, he was rescued from wasting a great deal of energy on self-torturing guilt—the kind of "perilous stuff that weighs upon the heart" and in the long run might well have given him real cardiac trouble.

In a lighter vein, the dreaming mind often uses puns based on body metaphors to bring a little humor into our lives when we need it. A good example was a friend of mine who dreamed that his left knee kept giving way beneath him as he walked. When we asked if he had behaved in a "weak-kneed" fashion recently, he laughed and said that he kept giving in to the demands of his ex-wife. This reminded him of an earlier dream in which he had been trying to throw a cat off his back; the more he struggled to get rid of it, the more it clung to him. "I've been trying to 'get her off my back' for a long time," he said, "and I thought being nice to her was the answer." The dreams made him realize that his "weak-kneed" attitude only increased the demands, and he resolved to be firmer in the hope of getting the "cat" off his back once and for all. This decision was promptly confirmed the following night by a dream in which he kept falling down under a heavy load, which he himself interpreted as meaning that he just "couldn't stand any more."

My dream collection includes many more dream cartoons based on figures of speech like being "two-faced," "starry-eyed," "level-headed," "lily-livered," "spineless," "hard-nosed," or having "butterflies in the stomach," or "one's arm twisted," "bending over backwards," "losing face," "keeping a stiff upper lip," "keeping an ear to the ground," doing something "behind one's back," and countless others.

FINDING PUNS IN YOUR OWN DREAMS

By this time you should have been able to spot some puns in your own dreams, but if not, don't be discouraged by the feeling that you are just not a "punny kind of person." We are all punsters at the back of the mind, and puns are bound to appear in your dreams sooner or later. When I lectured on dreams at a recent international conference, I was delighted when the chairwoman, Dr. Margaret Mead, expressed the hope that I would emphasize the importance of puns and body language in dreams. As a dream veteran who has studied her own dreams and those of people in many other cultures over the years, she too had been disappointed at the short shrift given to this aspect of the dreaming mind's creativity in the literature, in spite of the fact that most professionals are well acquainted with it in practice. I am reminded here of the late Dr. Fritz Perls, who, confronted with a girl's dream of a baby carriage running away into the traffic, asked sardonically, "And when did you have your miscarriage?"

So the next time you are looking over your dreams, just try repeating them aloud to see if any puns strike the ear—or imagine the dream as a cartoon and ask what caption it would have. Learning to spot dream puns is not only essential in understanding how the dreaming mind works, it also puts you in touch with basic resources of wit, creativity, and humor you probably never realized you possessed.

«7»

UNMASKING YOUR DREAM IMAGES

There would not *be* a dream from the unconscious except as the person is confronting some issue in his conscious life—some conflict, anxiety, bafflement, fork in the road, puzzle or situation of compelling curiosity. That is, the incentive for dreaming—what cues off my particular dream on a particular night—is my need to "make something" of the world I am living in at the moment.

—Rollo May

If you have been following the instructions so far, you should have at least a few recent dreams written down with notes of their *themes* (falling, being chased, meeting famous people, or whatever) and a record of any *dream puns* that have caught your eye. If you have also been able to relate your dreams to the events or thoughts of the day—without which any dream interpretation is incomplete—then several of your dreams should be clear to you. The majority of dreams, however, depict strange and even weird images and characters and usually do require further work before their meanings emerge. To tie the events and thoughts of the day to these dreams is particularly important because there are always several possible interpretations of each dream symbol, and only you can find the "correct" interpretation by relating it to something that was on your mind or in your heart as you fell asleep.

In deciding whether a dream image should be understood

literally or symbolically, the rules are:

1. If the dream character—human, animal, vegetable, or mineral—is a *real* person or thing in your life or on your mind at the time of the dream, then it should be considered literally in the first instance and taken symbolically when and only when a literal interpretation makes no sense. (Even Jung, the archexponent of elaborate dream symbolism, was insistent that dreams of a husband, wife, child, neighbor, colleague, the dog, and anyone with whom we are in intimate contact at the time of the dream almost always refer to the individuals themselves rather than to anything more subtle.) I know from my own experience that it is a mistake to interpret a dream of your car failing as a symbol of failing *drive* in yourself until you have checked the car, since the dream may well be throwing up subliminal perceptions of something wrong with the engine which you have been too busy to notice during the course of the day.

2. If a dream character or image cannot be taken literally as a real person or thing in your life, then it symbolizes either someone or something in your external life, or a part of your own personality which your heart is bringing to your attention. (Jung referred to the former as an *objective* interpretation, and to the latter as a *subjective* interpretation.)

In looking for the meaning of any symbolic dream image, always check first to discover whether or not it symbolizes someone or something external to yourself at the time of the dream, for we dream about the world outside us just as much as we dream of our private inner world. For example, if you dream of Vincent Price, and the real Vincent Price does not figure personally in your life at the moment, then look around to see if anyone else would fit the name. Is there perhaps a Mr. or Mrs. Price or a Vincent in your present life to whom the dream could refer? If not, then you must ask yourself what Vincent Price means to you. It could be something like costliness (a pun on his name) or showmanship (an association based on his qualities), or whatever else he may mean to you personally. Is there someone in your present life—husband, wife, colleague, neighbor, and so on—who has behaved in an extravagant (or showy or entertaining) way during the previous day or two? If there is, then your dream is probably expressing concern about your relationship with this particular person.

If you can think of no such person, then you have to consider the possibility that Vincent Price might be a part of yourself—which is nice if you admire him, and not so good if you dislike him! Have you behaved extravagantly or shown off during the past day or two? Always remember that the dream exaggerates in order to bring its point home to you; if you dream of a fascist and it turns out to represent part of your own personality, don't get too upset, for the dream is merely saying that you *feel* you behaved a bit like a fascist in the recent past, which may mean no more than some unpleasant thought about your Jewish neighbor or a dictatorial attitude toward your teenage son. If you continually dream of fascists and there are none in your life, either literally or figuratively, then it is probably fair to say that you have an inner conflict about this subject—but once again, you must remember that the dream merely reflects *your feeling* about yourself and your behavior, and your friends may not see you in this light at all. As Erich Fromm writes in his book *The Forgotten Language,* "Dreams are like a microscope through which we look at the hidden occurrences in our soul." In Part III of this book I discuss ways of dealing with some of these hidden occurrences.

In this chapter, I shall take several dreams from my collection to demonstrate in a practical way the various techniques you can use to discover the identities of your dream characters and images. I have chosen these few examples from thousands of dreams in order to stress what I consider the most important points in dream interpretation, but they cannot be more than guidelines at this point in the dream game. As I have said many times before, I cannot interpret your dreams: you must do it for yourself; my aim is to help you make a start. Even after applying all the rules and suggestions given in this book, you will almost certainly find that some dreams still elude you. Don't worry about this too much; it happens to all of us. But do continue to write down your dreams, together with the events and thoughts of the day, for many of them may become clearer as you get to know the meanings of certain recurring symbols over a series of dreams. Very often you will find that a certain elusive symbol in one dream reveals its meaning quite openly in another dream. When this starts happening, you are ready to compile your dream glossary, instructions for which are given in the Appendix.

The other principles to be borne in mind at this stage in the dream game are essentially those listed at the beginning of Chapter 5 in relation to the meaning of dream themes, but I shall expand them here in order to refresh your memory:

3. Even though dreams may take us back to childhood or concern themselves with future possibilities, they are always triggered by something on our minds or in our hearts at the time of the dream. People or things that were once very intimate parts of our lives—parents, siblings, childhood home, or friends—cannot be taken literally in our dreams if we are no longer directly involved with them. A dream does not indulge in reminiscence for its own sake. Such characters and images appear in our dreams either because they represent the voices of the past which still live on in us and influence our present behavior, or to tell us that something in our present situation reminds us of a similar situation in the past.

4. The feeling tone of a dream is always important and sometimes gives the clue to the meaning of a dream symbol.

For example, if I dream of a dog passing me in the street wagging its tail and feel very dejected in the dream because it does not respond to my friendly call, the clue to the dream's meaning may come in remembering how upset I was at my husband's behavior the previous evening at a "cocktail" party —and my dream could be reflecting my heart's thought that he was so concerned with playing the "gay dog" that he failed to pay me any attention.

5. If you happen to know from reading books that some particular symbol occurs commonly in people's dreams and has been stated by experts to have a universal meaning, by all means take this as a *suggestion* of what the symbol *might* mean if it occurs in your own dream, for our dreams pick up and utilize symbols from anywhere in order to make their point. Never assume that it must have this particular meaning, however, for there might be other more personal associations that are more important to you which actually determine how your dreaming mind uses this particular symbol. Since the majority of people in the West were brought up in a house, for example, a dream house is likely to mean "living space"—a symbol of your personality itself—but even this symbol can have different meanings in different circumstances. Always check what the symbol means to you—and always check on a possible literal meaning in the first instance.

6. If the same dream image or character recurs frequently in your dreams, then it is likely to have a similar meaning throughout a series of dreams, and for this reason it is helpful to compile a dream glossary of your own recurring symbols. (See the Appendix.) You should not be surprised to discover, however, that on occasion this particular symbol has a different meaning, and can sometimes be merely part of the background with no great significance. Any symbol is influenced to a great extent by the symbols it is grouped with in any one dream, and we should always see it

in the context of the dream as a whole.

7. Dreams do not come to tell us what we already know about the people in our lives or about ourselves, so if at first sight a dream seems to be doing no more than this, look deeper. At the very least, it may be clarifying the thoughts of the heart by putting them in vivid picture language or urging us to do something about a long-standing problem—but the dream may have an altogether deeper meaning which we can discover by looking again at its symbols.

8. A dream symbol is correctly interpreted when and only when it makes sense to the dreamer in terms of his present life situation and moves him to change his life constructively. Someone else may see a different possible interpretation, but this is only what your dream would mean *to him had he dreamed it.* Dreams do not arise arbitrarily from some universal reservoir: they arise out of the dreamer's present life experience and are meaningful to him alone. While I cannot say that the *purpose* of dreams is to move us to change our lives, I do insist that a "correct" interpretation—which means an effective interpretation—shows the way. For this reason, I suggest that anyone working on a dream successfully should conclude by writing down *briefly* what the dream means and *what he is going to do about it.* Jung made a habit of asking his patients, after they had worked together on a dream, "Now, *in one sentence,* what is the meaning of the dream?" We follow this rule in all our dream work, though we allow two or three sentences if necessary. And we conclude by asking the dreamer what practical action the dream message could lead him to take.

9. A dream is incorrectly (ineffectively) interpreted if the interpretation leaves the dreamer disappointed or diminished. Many psychotherapists still insist that they know the correct interpretation of your dream and believe that the message which makes sense to you may not be the one you need to see

at any given moment in time—apparently quite oblivious of the fact that their colleagues, on the basis of the same dream, may be seeing quite different things for you. You must learn to trust your *own* feelings and judgment.

The dreams that follow concern characters who are actual people intimately involved in the dreamer's present life, and they show the importance of taking these very seriously in a literal sense. I shall then go on to dreams involving symbolism, taking simple ones first, then more complex ones, indicating as I go how you can start to build a dream glossary of your own personal dream symbols and their various meanings.

NONSYMBOLIC DREAM CHARACTERS

While staying at the Association for Research and Enlightenment in Virginia Beach, I met Janis, who has been working for several years with dreams, following the teachings of Edgar Cayce and Jung. She told me a dream of the previous night, part of which depicted a man placing a necklace around her neck in a church. Later in the dream her husband, Fred, became infuriated at the sight of the necklace and ripped it off her neck, saying she could not wear it or even have it in her possession. When I asked what she made of the dream, she replied that she had not yet thought about it deeply, but she supposed that Fred symbolized the values she had "espoused," which were threatening some other aspect of her personality. Further conversation revealed, however, that the real Fred was very jealous of her for having the opportunity to devote full-time attention to spiritual pursuits while he had to stay at home earning the money, so I suggested that the dream might be more concerned with her literal marriage than with anything symbolic.

Janis wrote to me later confirming this: "I came down to the A.R.E. on September 16th and had the dream on the 20th. Fred was to join me the following day, so I didn't bother calling him. Meanwhile he was feeling very jealous and more angry with me as each day passed. When I talked with you about the dream, I had not heard that he was upset about my not having called: this only came out when I talked to him later about the dream and its jealousy theme. So I certainly feel now that Fred in the dream represents himself, and not a part of myself. As you said, your marriage can disintegrate while you try to understand what aspect of the 'inner husband' is jealous . . . certainly it is in some degree an aspect of your mind, or you would not be able to recognize or conceptualize it; but that does not mean it can't also be something else of greater *immediate* importance."

Jon, a member of one of our student groups, described a dream with a somewhat similar theme in his end-of-term paper, as follows:

> I am in a room at college talking to a friend. Looking out of the window, I notice that I can see across the court through the window into my second-story room. Sue is lying on the bed with Roger, a friend of mine. As I watch, Roger turns over on top of Sue and starts kissing her. I rush out of the room and run over to my room. When I get there, Sue and Roger are just lying on the bed talking.

As Sue was Jon's steady girlfriend at the time of the dream, and Roger a mutual friend in intimate daily contact with both of them, it was unlikely that these dream characters were anything but literal, so the dream almost certainly reflected Jon's feelings about these people. The day's events as described by Jon gave support to this notion: "Roger was over visiting us the night before I had this dream," he wrote, "I have felt that Roger's interest in Sue is more than friendly at times. He makes comments to Sue like, 'Why don't you leave that fag and come with a real man?' "

The dream's meaning is quite clear, and it only remained to find out whether it reflected an objective perception of the situation or merely neurotic paranoia on Jon's part. After discussing the dream with Sue, Jon wrote, "She agreed with me that Roger acts as if he is trying to hustle her. I think my subconscious picked up on this and caused this dream which reflects my worries about Roger's intentions."

An ideal analysis in Jon's dream diary would have appeared something like this:

Day's events and associations: Roger visited last night and made his usual verbal pass at Sue.

Dream theme: Girlfriend cheating with friend.

Dream characters and symbols: Sue and Roger — literal characters, themselves.

Dream message: I am worried about Roger's interest in Sue.

Action: Discuss with Sue.

This is a good example of the way dreams dramatize to make their point, for Sue assured Jon that there had been no actual cheating between her and Roger. The dream was warning Jon what *might* happen if things were allowed to continue. Many people dream of their partners having affairs with other people, and it is usually a sign that the relationship could be improved in some way. Very rarely is a dream partner symbolic. The same principle applies to neighbors and colleagues with whom we are in fairly intimate contact. I once dreamed of a new neighbor beating up his wife and was puzzled because he appeared to be a very kind person, but he did indeed prove on further acquaintance to be a man of vicious temper and primitive male chauvinist tendencies. Here too, the dream had dramatized its point—he does not actually beat his wife—but it showed that I had picked up real vibes from him without knowing it.

One of the most dramatic examples of the need to consider dream characters literally was given me by Rachel, an adherent of Gestalt therapy, which treats *all* dream symbols as parts of one's own personality and rejects the possibility of dream characters representing themselves. In this case the dream character was her own unborn child, and Rachel tells the story as follows:

> On the night we conceived the baby I, who rarely recall a dream, had the following experience. In my dream, I felt as if I were being awakened quite urgently, and the lights went out. I was then shown a slide show. There was only one picture in the slide show and that picture was of a fetus. I knew at the time that something was wrong with the fetus. I waited for another slide, but there was none. The lights went on and I slept the rest of the night. The dream came back to me the next day. As you may imagine, it was not a very pleasant memory.
>
> The dream recurred over the three months of my pregnancy approximately every five or six days. I would be awakened in my sleep, shown the slide and that was all. The dream was vague to me in the morning, but rendered a shadowy feeling to my day. After close to three months of this experience, which I alternately described as neurotic behavior and believed was an important message, I came to the conclusion one night that "something is basically wrong here, I don't know if I am ready to have a baby. I want to do some work on this." It was about midnight when we went to close friends (the friends being therapists) and I announced: "I am very upset. I don't think I'm ready to have a baby." My friend, who knows me quite well, said, "Rachel, you are being impatient again, you don't get a baby right away. You don't go downtown and get a baby. You have to wait nine months. There is nothing wrong with you."

> I felt very relieved after this brief talk, which ended with a salami sandwich at 1 A.M. and merrily went to bed to once again have the slide presentation dream. The next day I started cramping and the following day I had a miscarriage. . . . the results of the D-and-C showed that the fetus had lodged in the wrong place from the beginning.

Rachel's dreaming mind was clearly throwing up her body's awareness of the fetus's condition, which her waking consciousness could not detect.

On a few rare occasions, however, a person or thing which forms an intimate part of your life can take on a symbolic meaning in dreams. Fiona, my six-year-old daughter, often appears as herself in my dreams but also symbolically as "my baby"—that is, my dream work—and as the "natural child" of my own personality. She can also appear as some kind of small animal, depending on how I feel about her at the time of the dream. The golden rule is not to consider a dream character symbolically until you have ruled out the possibility of its literally being itself. When you are really satisfied that you have thoroughly checked this out and eliminated it, then and only then can you legitimately move on to a symbolic interpretation.

SYMBOLIC DREAM CHARACTERS AND IMAGES

Jim, who attended one of our British workshops, dreamed:

> I am walking down a village street along which tradesmen are displaying their wares. In addition to the greengrocer, butcher and so on, there is a very respectable tailor hanging out his suits on a rack in front of his shop, and I notice with some surprise that he is wearing a clerical collar. As I pass by, I suddenly realize that the whole street is a fake, a facade for a secret poison gas factory—and at this moment, my arms are

suddenly pinioned from behind and I wake up in panic as I know they are going to kill me for discovering their secret.

At first glance nothing resonated at all. Jim had not been to an open-air market the previous day, nor had he visited his tailor. A great part of the day had, in fact, been spent with his lawyer in a vain attempt to wind up Jim's divorce proceedings, which had now entered their fourth year of unpleasantness and distress for all concerned. So the day's events seemed to offer no clue to the dream's meaning.

Associating to the tailor, however, Jim was reminded of a tailor who had been a great authority figure in his childhood. He said his name was Ash—and then realized with a start that this was his lawyer's name. With this revelation, Jim then saw the obvious pun—the lawyer was "hanging out the suit," which, of course, referred directly to Jim's protracted divorce case. This was very interesting as Jim had been blaming his wife for the delay, whereas his heart evidently saw the situation quite differently: the lawyer himself was responsible for hanging out the suit on account of religious prejudices, as depicted in the dream by the clergyman's collar. Once Jim thought about this, it resonated totally with his bones, and as he kicked himself for having been so blind, he realized that the dream had also shown the reason for his blindness, by depicting Mr. Ash not as a lawyer but as an authority figure of childhood. It was Jim's awe of such authority figures that had prevented his questioning the actions of his lawyer. However, the poison gas factory remained unexplained until a friend of Jim's remarked several weeks later that this particular lawyer had poisoned her entire family life by his sanctimonious attitude to separation and divorce before her husband died.

Jim completed this page in his dream diary as follows:

Day's events and associations: Visit to lawyer about divorce.

Dream theme: Discovery of a dangerous situation.

Puns: "Hanging out the suit."

Dream characters and symbols:

Mr. Ash, the tailor (a past authority figure)	—	Mr. Ash, the lawyer (a present authority figure)
Clergyman's collar	—	moralistic attitude
Poison gas factory	—	something poisonous about the situation
Butcher and grocer	—	props in the drama: no special significance

Dream message: I feel that Mr. Ash is hanging out my divorce suit because of religious prejudices, and his respectability is a facade to cover a really poisonous situation. My reluctance to challenge him sooner stems from my fear of authority figures.

Action: Change lawyers or take control of my divorce proceedings and complete them as soon as possible. In either event, never employ Mr. Ash again. On a long-term basis, liberate myself from fear of authority figures.

John, my husband dreamed:

I have been invited to attend an international banquet at the White House, and the ceremony begins with President Nixon's own party standing round him singing, 'Hail to the Chief,' while the President puts on an air of mock modesty. Then as we walk down the corridor to the banqueting room, my arm is grasped by the President who gasps, "You've got to help me, John. Stand in front of me, please, and make sure the others don't see . . .," whereupon he proceeds to blow his nose loudly and messily on the curtains. Out of the corner of my eye, I see Premier Chou En-lai of the People's Republic of China and I think to myself that whatever mistakes Nixon may have made, he will always be remembered as the great integrator of East and West.

The events of the previous day had been nothing out of the ordinary—in fact, John had not ventured outside the house, nor had any visitors called. We had spent the day working on the book and had ended with a brief row, but nothing serious.

As John had not actually been invited to the White House, and as President Nixon did not figure literally in his life at the time, John had to ask whether Nixon stood for someone or something in his present life or as a part of his own personality. John's first association was that our house had just been painted white, so the dream drama obviously referred to something that had taken place within our own "white house." As we discussed the dream over breakfast, I gave a loud sneeze and blew my nose, which reminded John that I had, in fact, started a cold the previous day. There was no getting away from it—Nixon symbolized me! Having appeared as the Queen several times in John's dreams, I had to accept the possibility, but Nixon, I thought, was going a bit too far! Besides, I have never blown my nose on the curtains.

But by this time the whole dream was falling into place. John recalled how I had gone through some old mail the previous day in an effort to clear out some papers and had read out to him some fan letters before finally disposing of them. We had then worked and rounded off the day with a row in which, I admit, I behaved with less than dignity. John's dream reflected his feeling about the day in beautiful picture language, by depicting "the President" surrounded by adoring fans all singing "Hail to the Chief," and then disgracing "himself" where only John could see. Obviously his waking mind had turned over, without his becoming consciously aware of it, the fantasy of what my fans would think if they knew what I was really like. "Don't tell them, John," he imagines me begging, "help me, please." But as he wallows in this secret satisfaction, he notices Chou En-lai, which reminds him that whatever mess I may make in private life, he still sees me as the great "integrator" of opposite viewpoints and conflicting theories in the world of dreams. I appreciated this very much because now, whenever I am less than perfect in my daily life, I can say to John, "Remember Chou En-lai!"

John completed this dream in his diary as follows:

Day's events and associations: Outside of our house was painted white. Ann read out fan mail, and we worked on the book. We had a row.

Dream theme: Meeting and interacting with famous people.

Dream characters and symbols:
The White House — our house
President Nixon — Chief of our white house—Ann
The Chief with a cold — Ann with a cold
Aides — fans
Curtains — self-image, persona
Blowing nose on curtains — loss of face
Chou En-lai — dream expert with opposing views

Dream message: I feel that Ann loves admiration, but doesn't want the world to see her "Watergates" as I do. But I feel she is a real pioneer, so I can't put her down too much.

Action: In future, express feelings of annoyance and irritation, instead of waiting for them to come out in a dream.

I am going to digress here for a moment to point out that all dreams depicting famous people or royalty have meaning in terms of one's present life and, like any other dream symbol, they represent either someone else or a part of ourselves. For example, in the early days of my first marriage, I had recurrent dreams of the British Queen Mother dropping in unannounced at all hours of the day to give me the latest recipe for a cheap beef dish or the current market prices of meat and vegetables. In one dream she arrived as usual and proceeded to teach me how to make chicken soup, despite my protestations that we were both on a reducing diet. Each time I would wake up crying with anger and frustration, for how could I be overtly rude to the Queen Mother?

Although I was unsophisticated in the art of dream interpretation at the time, I soon noted that these dreams occurred after one of my mother-in-law's visits, and I had little difficulty in seeing that she was the Queen Mother of my life at the time, the royal ruler of my husband's household who was determined to train me in the art of homemaking, even though she knew quite well that I was studying to be a doctor. The dreams ceased when I put my foot down and made firm visiting hours for the "Queen of the family," who never forgave me for questioning her authority. Since our divorce she has reclaimed her place once more as Queen Mother and rules my husband's second family in the way to which she had formerly been accustomed.

I am reminded here of the large number of women patients in my first husband's practice who reported pleasant and often erotic dreams of Prince Philip. It did not dawn on me at the time that they were probably dreaming about my husband, whose name was Philip, and that their dreams chose to depict him in this way because they saw him as a very handsome, princely person—a true son of the Queen Mother!

Psychologists and laymen alike tend to dismiss dreams of mixing with famous people as mere wish fulfillments or compensation for the dullness of ordinary life. Typical of this point of view is a book by Brian Masters, a British author, entitled *Dreams of Her Majesty the Queen and Other Members of the Royal Family,* which reports hosts of dreams from people who never came near the royal family in waking life. Many of the dreams are hilarious: in one, the Queen is keeping horses in her bedroom because the royal stables are gone ("They said it was unhygienic, but I still have a little influence," she said); and another depicts the Queen standing under Niagara Falls muttering something about having to wear "this ghastly mackintosh," at the same time trying in vain to hold her tarnished crown on top of her rainhat because "one's people expect it of one over here." Housewives by the score dream of the Queen dropping in for tea: in one dream she is very embarrassed at bringing an uninvited guest, and there behind the door, wearing gum boots and looking very sheepish, is the Queen Mother. Disappointingly, the only significance Masters gives these dreams is that they bring some color into the dreamer's gray existence. "You dream of royalty because exalted, regal figures compensate for the dullness of routine existence," he writes. "They inject some excitement into your prosaic world." I would make a bet with Masters that if any one of his dreamers had presented me with a dream of this kind, along with the day's events and associations, we would have been able to identify the royal figures either as people in his life or as aspects of his own personality. Dreams do not come merely to entertain us, though they are often entertaining. They come to move us forward to a new level of understanding of others and ourselves.

Mary, who helped initiate a small dream group among friends, reported the following dream:

> I am in a vacation place and there is something wrong with my stove. I am thinking I need to get it fixed. There is an explosion and part of the top of the stove blows off and there is a puddle of water or gas on the floor. I feel I must do something but I can't bear to look at this stuff. Smoke is coming out of the burners and it looks as though a hole has been burned through the fireplace. I think I must send for the power company. The ashes are still smoking.

Following our rules, Mary checked out her stove to discover whether or not the dream might be warning her of a real fault which could lead to a dangerous explosion. Finding nothing wrong, she then asked herself what "stove" meant to her. The best she could manage was that it is a piece of kitchen equipment used for cooking food by means of heat. Now, whenever a dreamer finds it difficult to associate meaningfully to a dream character or image, we ask him to try the technique of "creative monologue," which consists of "becoming" that particular dream image and describing yourself. So Mary sat in her chair, closed her eyes, and tried to get into the feel of what it was like to be a stove. Then giving the stove a voice, she began: "I am a stove, a piece of kitchen furniture. My place is in the kitchen. People switch me on when they want to use me, and switch me off when they have finished with me . . . I'm fed up with being treated like this . . . I'm used and neglected . . . and if someone doesn't pay attention to me soon, I shall explode."

The creative monologue technique can be used whenever a dreamer finds it difficult to associate meaningfully to a dream image, and it does not always lead to the conclusion that the image is some aspect of the dreamer himself. For example, if Mary's husband had dreamed of a stove exploding, he could have used the technique to discover the stove's identity, and in so doing also gain some insight into the explosive feelings of his neglected wife. It is a very powerful technique, used constantly in our dream groups to identify elusive dream symbols, and is particularly useful in enabling group members to see things the dreamer himself may have missed, for very often aspects of the dreamer's personality unknown to himself emerge as he acts the part. As I have already stressed, suggestions and comments are always welcomed from the group, but they must resonate with the dreamer's bones to be truly useful to him.

Having identified the stove as the part of herself in danger of exploding, Mary looked back on the events of the previous day to discover the motivation for her dream. She writes:

> When Jane, Alice and I met on February 28th, it was to work on some of my dreams that I didn't feel free to work on in the group with my husband. I had gotten up in the morning feeling angry, but could not account for why—I suspected another dream of being a maid in my own household, but couldn't recall anything. Jane was in a terrible state when she arrived, and I soon forgot my angry feelings and became absorbed in a real desire to help her so she could get through the next three nights at the club. When I listened to the tapes later that afternoon I started feeling absolutely awful about how much aggression I heard in my voice and the way I prodded Jane to stand up to the punishing voices inside her. Not once did I have any feeling of being angry with Jane, nor was I, but I could not account for the way my manner and voice on the tape made me feel. Although Alice expressed the feeling that indeed there was aggression on my part, which seemed necessary to help Jane at this point, I kept sounding to myself like a fishwife. After Jane left, I spent another three hours with Alice listening to some bad feelings she was having about herself concerning a situation with her child. At the end of the day, I was still sitting on my angry feelings and neither I nor anybody else had dealt with them. That night I had the stove dream.
>
> My interpretation of the dream is this. I am, of course, the stove, and the Gas Company (Power Company) is the Dream Power Company, the group I called in order to discuss my miserable dreams of being a maid. I had felt unable to discuss my feelings of being used and neglected by my husband in his presence and had decided to call an extra group while he was away. So the dream first of all tells me how I handled the day's events: I had called together a meeting of the company to discuss my dreams, for I realized there was something wrong with a part of me (the stove). Jane's and Alice's problems seemed more pressing than mine so I did not look at my need for attention, but instead helped them deal with theirs. When

later I looked at my feelings (the ashes), I was burned up and still smoking, for I had exploded, pushed the anger (ashes) down to a lower level where it was still smoking and burning a hole in my gut. It was clearly I who did not communicate with the power company that I was about to explode if I did not pay attention to my anger—and this was anger I had brought into the group and not anger I was feeling toward Jane and Alice.

The dream is also a statement about my life generally. My conscience is always barking at me that my needs are at the bottom of the list in priority. The message to me in the dream is that I should listen to my feelings and give them a voice, else I will explode and cause damage. I did not understand what the puddle on the floor was in the dream until Jane was here one day when I was at the bottom of the barrel and in great need of someone simply to hold me, and when she put her arms round me I immediately burst into a flood of tears. My nature is such that I relieve pent-up emotion either by an outburst of anger or a flood of tears—and sometimes both together.

Good-bye—I must go and clean my stove!

Mary

Here is Mary's dream as it appears in her dream diary:

Day's events and associations: Called dream group to discuss my dreams, but the others' problems were so pressing that we never got round to mine.

Dream theme: Explosion.

Puns: The stove "blows its top."

Dream characters and symbols:
Stove — part of myself in danger of exploding because it feels itself used but neglected
Power Company — Dream Power, i.e. the dream group
Puddle — tears
Ashes — burned-out anger
Fireplace — my gut

Dream message: I must express my feelings when they arise (if possible) and demand attention for my needs, otherwise I shall explode into anger or tears, or both.

Action: Keep in touch with my feelings at all times.

Barbara, a college teacher from Georgia, dreamed three dreams in one night, which she describes as follows:

1. I see three fields—a small one, a medium-sized one, and a large one. They are somehow symbolic of sex (when I awoke, I had the image of the fields in my mind and knew I had been dreaming about sex). I returned to sleep.
2. I am on a basketball team (a recurring motif). I am working out with a small group, wondering whether I will have the energy to keep up with them. I think so. Will I be a "forward" or a "guard"? I am shorter than several other players, which might inhibit my style as a guard, but I don't know if I can shoot well enough to play forward.
3. Malcolm R. and his friend, Dawn, are visiting us. There is a feeling of relaxed friendliness. I am showing Dawn the clothes hanging in my closet. Without offending me, she discreetly suggests improvements that need to be made in my wardrobe—mainly that I sew some buttons on my jeans. Although Charlie and I live in the house, I am relatively unfamiliar with it. I walk among the rooms the next morning and am surprised to find the bedrooms adequate for several guests to stay overnight. They are spacious and clean. I notice the bed that Malcolm and Dawn slept on. It has a tall platform on which the mattress rests. It looks heavy and solid.

At first glance nothing clicked at all with any of these dreams, nor did they appear connected in any way. There were no literal fields or basketball games in Barbara's life at the time, nor had she seen Malcolm and Dawn lately; moreover, she is always meticulous about her clothes and her jeans were in no need of repair. She recalled how the previous day she had discussed her ambivalence about motherhood with some friends and had fallen asleep still thinking about this. At the time of her dream Barbara was in her early thirties and childless.

She decided to tackle these dreams in order but could get no lead into the first fragment at all, so she applied the creative monologue technique and "became" one of the fields. "I am a field," she began, giving it a voice, "and I am about to be planted . . ." In just a few words it had become apparent that the field was a womb symbol and that Barbara's heart felt she was ready for pregnancy. This immediately suggested the possibility that the field upon which Barbara and the team were playing in the second dream carried the same meaning, and with the recognition of several puns the meaning of the dream quickly became clear. The dream depicts Barbara in the process of "balling" or having sexual intercourse, and wondering whether to go "forward" into pregnancy or "guard" against it.

The third dream, which at first sight appears to be quite unconnected to the others, contains a pun that places it in the context of the night's dreams as a whole. Dawn points out that Barbara's jeans (genes) are in poor condition—a beautiful dream cartoon expressing Barbara's doubts about her heredity. Barbara wrote, "Here are some definite connections. I had a few weeks earlier dreamed of my younger brother as a baby-killer, and in that dream I accused my mother of being the cause of his mental troubles. The present dream seems to show unconscious concern on my part about passing on these problems to my offspring."

Continuing with the dream, Barbara remarked that as she had not seen Dawn for some time, she would have to understand her as a symbolic character, so she asked herself what Dawn stood for in her mind. The answer was "refinement," and Barbara wrote that "Dawn grew up as the daughter of a well-to-do landowner and has a finishing-school education. She is an artist of refined sensibilities, and she is also at one with her body—she rides horses, swims, practices yoga. She seems always to have everything 'together,' including her clothes. In contrast, I see myself as constantly struggling to get myself together, including my clothes (a recurring motif in my dreams). Malcolm is perhaps even more refined, and they are both extremely good-looking."

Barbara concluded that "the dream shows my misgivings about my completeness as a woman. I grew up in circumstances of economic and social stress and spent my life consciously trying to overcome my disadvantage. Dawn is my opposite in this respect. Will my lack of social and particularly feminine refinement impair my ability to experience the sexual fulfillment of motherhood? But I am shown that my house (which I always associate with my Self and my potentialities) is larger than I realize and that there is plenty of room for guests (babies?). Moreover, the dream reassures me that my *back-ground,* symbolized by the bed, far from being a deficiency, forms the solid and stable basis for my future growth."

These dreams appear in Barbara's dream diary as follows:

Day's events and associations: Had spent the previous day at the K's. In the evening we discussed my feelings about my family and the possibility of my having a baby.

Dream themes:
1. Sex (field)
2. Playing basketball (recurring motif)
3. Something wrong with my clothes (recurring motif)
4. House larger than I expected (recurring motif)

Puns:
"Forward" — to go ahead with something
"Guard" — to be on one's guard against something
"Basket-*ball*" — sexual intercourse

"Jeans" — genes
"Bed" — *back*-ground

Dream characters and symbols:
Field — womb
Malcolm and Dawn — social refinement
House — my Self, psychic space, potentialities

Dream message: My body feels ready for pregnancy, but I am of two minds about it on account of my "poor genes" and lack of refinement. But my heart assures me that I can do it successfully, and that my background, far from being an obstacle to growth, forms a solid basis for further growth.

Action: Go ahead and have a baby!

And so she did. A few weeks after the dream Barbara wrote to say that she was pregnant and was delighted about it. In another letter after the birth of her son she wrote: "Charlie and I have never 'planned' a baby, nor have we gone to extreme measures (i.e. the pill) to avoid having a baby. It seemed to be something that would happen in its own good time when we were ready for it. I believe Jonathan happened that way."

Margaret, a friend with whom we stayed in California, wrote to us soon after we left to tell us she had dreamed about me and was delighted to discover that she could interpret the dream by herself. In a letter she described the dream as follows:

> Ann [Faraday] is engaged to be married to a friend of mine, Fred. I figure only as an onlooker and can see that Fred is really getting on her nerves no end. I like both of them, and am sorry to see that his behavior is going to result in his losing her. As evening approaches, he suggests she go to bed early as she has had a tiring day; at mealtimes, he presses her to eat more as she is losing weight; and as she goes out, he gets a sweater and puts it round her shoulders—the whole Jewish

> mother act. Ann in the dream is just as she is in real life—perfectly able to look after herself and make her own decisions, and Fred is like the real Fred—a very kind guy who really needs looking after much more than she does.

Speaking directly to me in the letter, Margaret described how she worked on the dream:

> The dream seems to have nothing to do with me at all, and obviously has nothing to do with you and Fred because I haven't seen either of you for a long time. I was beginning to think that I wouldn't be able to interpret my dreams now you had gone—and then the idea came to "borrow" you wherever you were. So I conjured you up and sat you in the car beside me—and believe it or not, "you" came up with all the right questions. In the first place, you told me to ask Fred why he fussed Ann so much, and "he" answered, "It's because I love her so much, and that's the only way I can express the love I feel for her all the time."
>
> To cut a long story short, I then put myself in "Ann's" place, and immediately got a clear picture of my mother fussing me. Even now when I visit her, she will call to me as I go into the dining room "come and have breakfast," and just as I am about to sit down, she will say, "sit down and eat some breakfast." If I plan to drive to town, my father will back the car out of the garage for me, etc. etc. etc. I suppose at the back of my mind somewhere I knew all this fussing over small details was an expression of love, but the knowledge was well covered by the annoyance of it all, and the resentment that they would do all the easy things for me and none of the things that really count. It was just a lot of talk, but looking at the dream, I see they really don't know any other way to express their love, and I really felt my mother's hurt confusion at having the fussing (i.e. love) rejected.
>
> This put me in touch with my own current feelings of loss—of leaving my daughter, Deborah, here in California while I go back to Washington, and I realized that I had been buying

her new clothes, seeing she had enough underwear, etc., and I saw that I was doing exactly what Fred in the dream and my mother in reality would do. So I told Deborah the dream, and we laughed over the substitution of underpants for straightforward expression of love.

The completed dream appears in Margaret's dream diary as follows:

Day's events and associations: Planned my return to Washington. Bought underwear for Deborah.

Dream theme: Jewish mother act.

Dream characters and symbols:
Ann — Deborah in her self-sufficient aspect
Fred — myself in overprotective aspect

Dream message: My fussing Deborah is a real expression of my love and concern for her but will spoil our relationship by irritating her.

Action: Tell Deborah the dream and try to express my feelings openly. In the long term I should get in touch with my own need to be looked after sometimes.

Had the dream been making the simple point that Margaret was being overly protective toward her daughter, then it would probably have depicted Margaret herself fussing over Deborah. But it obviously had much more to say than that. By using Fred the dream made the additional point that here was a person who needed love, care, and attention himself, and by using the image of Ann, Margaret was able to get the message that Deborah was a big girl now, quite capable of looking after herself, which made her feel much happier about leaving her behind.

Margaret's technique of overcoming her initial sense of helplessness in the face of this dream was an interesting one: by conjuring up the mental image of me telling her what to do, she got in touch with the fact that she actually had the knowledge of what to do inside herself. Although I am sure Margaret did not know, this is a technique used by many Jungians when they get stuck—by conjuring up the figure of Jung himself and giving him a voice, they are able to tap their own inner resources. And Fritz Perls actually suggested that "Fritz dolls" should be marketed for the same purpose. So I shall be pleased to make myself available—at the breakfast table, on the plane, or wherever you happen to be—if you still need help with your dreams after reading this book.

THE THREE FACES OF DREAMING

When you become practiced in the art of dream interpretation, you will notice that your dreams find their most effective meaning at one of three energy levels. In *Dream Power,* I called these the Three Faces of Dreaming, and I will illustrate my point by showing how all the dreams in this chapter found their effective interpretation at one or more of these levels.

Looking Outward, we find that many dreams provide valid information about people or situations in the external world. Such dreams are usually triggered by subliminal perceptions or vibes picked up during the day but never consciously registered by the waking mind. For example, Janis had picked up real vibes of jealousy from Fred; Jon had detected Roger's more-than-friendly intentions toward his girlfriend; and Jim correctly perceived that his lawyer was holding up his divorce on moralistic grounds, as verified by a friend at a later date. Dreams that give their most useful message at this level are usually warnings and reminders that we cannot afford to ignore, and very often immediate action is necessary. For this reason, we never go into more complex interpretations before checking out the external validity of our dreams.

Through the Looking Glass of dreams we receive messages about our *subjective* reactions to the people and situations in the external world. All dreams are in a way pictures of subjective reality, but some have more objective validity than others, and it is up to the dreamer to decide where the energy of his dream lies. John only *felt* that I had behaved like Nixon. My first husband's patients *saw* him as Prince Philip, and I had an *inner image* of his mother as the Queen Mother. Such dreams reflect our own private feelings about someone or something in our lives at the time of the dream, maybe shared by others, maybe not. Similarly, Margaret's dream merely reflected her *feeling* that she was behaving like Fred and that Deborah was like me—feelings that may or may not be objectively true. Looking-glass dreams, unlike those that look outward and find their energy almost entirely on an objective level, seem to derive their energy in equal amounts from inner and outer worlds, so that objective reality becomes somewhat obscured by subjective feelings.

Looking Inward, dreams give us a picture of how we feel about our own private inner world, as illustrated by Mary's dream of the exploding stove. The stove was not a symbol for someone or something in her external world but represented a part of herself, as did Dawn in Barbara's dream of the "poor genes" and Fred in Margaret's dreams. Both these dreams concern conflicts, not with characters in the external world but with split-off portions of the self within the depths of the psyche. Part III of this book is concerned mainly with looking inward and finding ways of resolving inner conflicts, getting alienated parts of the personality "together," and becoming a whole person.

The aim of the dream game is to look for *whatever* messages your dreams have to give you on *any* of these energy levels, and of course you can decide only on completion of your interpretation. In some cases you will find that a dream has more than one message for you—as indeed Margaret's dream did, turning out to be concerned both with her relationship to her daughter and also to the "daughter within." The energy level of the dream determines what action you take in heeding the dream's message, so it is important that you make your own assessment and not allow yourself to be swayed by someone else's interpretation. If you have any feeling about what the characters in your dreams do or do not mean, *trust that feeling* and not any "expert's" interpretation.

If the meaning continues to elude you, take Jung's advice and "Look at it from all sides, take it in your hand, carry it about with you, let your imagination play round it and talk about it with other people . . . treated in this way, the dream suggests all manner of ideas and associations which lead us closer to its meaning." He adds reassuringly, "I know that if we meditate on a dream sufficiently long and thoroughly . . . something almost always comes of it."

One quite effective technique is to sit or lie down comfortably during the course of the day and ask that an image come to mind which will give you a clue to the meaning of your dream. Turn the dream over slowly in all its details, repeat the process and allow yourself to drift, keeping yourself sufficiently receptive to any images that arise. You can also do this in the middle of the night, or before rising in the morning, if the dream you have just had is not clear to you. Often the image will be a face, which may give the clue to the identity of a particular dream character, or it may be a memory flashback reminding you of a similar situation in the past. It does not always work, but it is worth trying.

If all efforts fail, then you can try a new experiment that works well, and ask your dreams for help. Before falling asleep you can ask your dreams to give you the meaning of a certain elusive symbol, or even to replay a puzzling dream in terms you can understand. We have found this technique so effective that we also use it to get our hearts' thoughts on waking problems of work, relationship, or whatever, and on new life projects and endeavors.

«8»

ASKING YOUR DREAMS FOR HELP

> Ask and it shall be given you.
> —St. Matthew

Since the main problem in understanding a dream is to discover what issue on your mind or in your heart provoked the dream, you can take a shortcut by asking your dreams for help on a certain problem of emotional significance before falling asleep. One issue on which you can be sure your heart is involved is a dream you have been unable to understand, as it is your heart that it came from originally. Every dreamer must find his own best way of eliciting help from dreams. I like to call on "dream power," and when I am fully relaxed in bed with the lights out, I put my request something like this:

> Thank you, dream power, for the dream of the island (or whatever) you sent last night. I feel it's an important dream, but despite all my efforts, I'm still unable to get the message. Please send me another dream tonight putting the same message in a way I can understand. I'll make an extra effort to remember my dreams, and I promise to write down anything you send immediately on waking. Thanks.

I fall asleep repeating this request, being careful to emphasize that I will recall my dreams and write them down, adding the thanks as a gesture of confidence in dream power. Some people may prefer a firmer approach like autosuggestion, and it is perfectly good practice to fall asleep repeating something like, "I shall remember a dream tonight (or in the morning) which will clarify last night's dream of the island. I shall remember my dreams and write them down on waking." Religious people to whom prayer comes naturally may like to ask God for enlightenment on the dream. However you frame your request, it is essential to have your recording equipment ready, since failure to do so is a sure sign that you are not serious, and the unconscious mind is not fooled.

DREAM POWER AS DREAM INTERPRETER

Very often the dream you get in answer to your request for clarification seems nothing like the original dream at all, but a little work will soon show that the basic message is the same. A classic example of this was provided by John, with the solution of a dream mystery that had all the excitement of a detective story.

The mystery began when he had two dreams on the same night with the common theme of finding unexpected value in old furniture inherited from his working-class father. In the first dream an old sofa from his childhood home turned out to be a valuable antique. In the second he was about to junk an old pedal organ which his father had (in real life) bought for a few shillings to play in the evenings, and had later converted into a desk for John's first apartment. Just before he let it go in the dream, John decided to check behind the boards to see if any papers had fallen down there and was astonished to find a large sum of money.

The repeated theme suggested an important message, and there was no question of a literal meaning, since both the sofa and the organ/desk had long since fallen to pieces. It seemed, therefore, that John's heart was urging him to recognize some *values* he had inherited from his father which neither of them had appreciated, and this was confirmed when we tried the creative monologue technique. Taking the role of the sofa, John said, "You've all been *sitting on me* for years without seeing my worth." The trouble was that he simply couldn't imagine what this value might be, and neither the dream nor further dialogue brought the vital clue. So he asked for a clarifying dream.

At first sight, the resulting dream seemed to have nothing whatsoever to do with the request. It showed John climbing up the outside of the building where he had worked for over twenty years before leaving to become a free-lance writer. Feeling very frightened in the dream as he climbed up, he decided to go in through a window and use the internal stairs instead. He found himself in a room with a locked door. As there was no one else in the building, and no possibility of getting out of the window again, he awoke in a panic of claustrophobia.

Working on the dream the next day, John acted his own part in the drama and said, "It's scary trying to make your way up on the *outside,* but it's far worse on the *inside* because you feel completely trapped." John resonated to the fact that this had been his feeling about leaving the big company that had sheltered him all those years—and suddenly the earlier dreams fell into place.

All his life John's father had been poorer than his friends thought he need have been, because he absolutely insisted on remaining a self-employed worker, which for a plumber in Britain of the 1930s meant a very hard life. As he constantly complained about his poverty, John took him at his word and sought economic security by joining industry. When he finally left to make his way on the *outside,* he had felt distinctly scared and from time to time when panic overtook him wondered about taking another *inside* job. The dream was telling him in no uncertain terms that this would bring a feeling of claustrophobia far more frightening than the insecurity of making it alone in free-lance work. Finding money in the dream desk also enabled John to reclaim his own long-lost hidden resources and confidence in his ability to survive on the outside. What John had considered a neurosis in his father—the refusal to work for someone else—turned out to be a liberating value which, if recognized sooner, could have spared John many long years of imprisonment in industry.

After using this method of dream clarification for several years in both individual and group work, I discovered that Edgar Cayce had made the same discovery many years earlier through his trance work, quite independently of any professional psychological expertise. He added a suggestion that I found particularly intriguing, namely that the request for a clarifying dream would sometimes be met by a special kind of dream which he called an "essay"—a picture drama accompanied by an explanation given in a verbal commentary by what he called the "interpreter" or "interviewer." I was inclined to be skeptical of this idea, since all the evidence suggests that verbal statements are not the normal language of dreams. I decided to try the experiment, however, and deliberately asked dream power for such an essay to clarify a certain symbol that had been recurring recently in my dreams—the symbol of a plane in which I sometimes travel, but often await in vain and even miss.

My most recent dream of this kind, which occurred four days before I put my request, found me waiting in a field for a plane that didn't come. As I waited, I saw an old-fashioned train grinding slowly across the field and realized that I must be in the wrong place altogether, since no plane could possibly land there. I set out to look for the airport which I knew could not be far away, as the setting was a small island. Working on the dream the next day, I guessed that the train was probably a pun symbol for "training"—a metaphor I had lifted from John's dreams and used several times in my own—but the meaning of the dream as a whole remained obscure until I asked for a clarifying "essay" and the "interviewer" provided the answer.

In a very vivid dream I found myself in a TV studio about to repeat an interview in which I had participated some days before. A somewhat harassed woman interviewer dashed in, sat down opposite me, and asked, "Now what was our subject —was it astral projection and out-of-the-body experiences?" I replied, with a laugh, "I hope so because that's what I talked about." And with that I woke up.

I was astonished—and also highly amused—that dream power had taken Cayce literally and provided me with an "interviewer" in the form I was most familiar with in waking life, yet at the same time had brought me the clarification I sought. Her question told me that the plane in my earlier dreams was a pun symbol for the "astral plane," the reference being to my recent experiences in which the dream body (often called an "astral *vehicle*" in occult literature) seems to leave the physical body and travel in a world far more real and vivid than either dreaming or waking. The dream of waiting for the plane on the island had occurred at a time when my "astral vehicle" had not put in an appearance for a while, and I had been turning over in my mind whether I should intensify my yoga and meditation in pursuit of heightened consciousness. The dream gave me my heart's answer, namely that I saw this as a terribly slow, grinding "training" that would take a lifetime to get anywhere and would not lead to the kind of experience I was after at the time.

Cayce suggested that we should ask for clarification immediately after the puzzling dream, either during the night if we awaken with such a dream in mind, or the following night if this is not possible. This may explain why my "interviewer" appeared harassed, as I was asking for a repeat performance of something that happened several days earlier. "Look, I do five programs or more every night," she said in effect, "so it's a bit much expecting me to remember one I did so long ago." However, dream power obviously did remember and obligingly provided the answer, but I took her remark as a slight reprimand not to overwork her in future. I have tried the experiment of asking for an essay complete with verbal interpretation since but without much success, which corroborates my feeling that it puts a strain on dream power, whose natural language is that of pictures. The experiments, however, may be biased by my expectation, and I should be very interested to hear from readers who often have clarifying essays in the form described by Cayce.

I would certainly agree with Cayce that if you are going to ask dream power for help in understanding a dream, it makes sense to do it soon. Never make more than one request a night and keep it as simple as possible. For this reason, I recommend sometimes confining requests for clarification to puzzling symbols rather than to whole dreams—though we have had some remarkable results from the latter. When one of our students dreamed of killing a wolf and asked for clarification of this symbol rather than the whole dream, she dreamed she was standing over her father's grave. With the realization that the wolf was her father, the rest of the dream's meaning fell into place.

It is rare, however, to find that all the pieces of a dream fit together like a jigsaw puzzle, and it is a waste of good dream time and energy to ask that every single item make complete sense. I used to do this in my early days of dream research, when I was fired with the miracles performed by dreams, until dream power in its wisdom brought me a sense of proportion.

I had asked for help in unraveling the meaning of certain minor elements in a dream, and was shown a tree. My feeling in the dream itself was that a dream is like a tree whose roots must remain underground if it is to grow and produce fruit. On waking, I took the point—dreams are like life, and to try to decode every tiny element is like trying to freeze a living process. Probably the web of associations and thoughts out of which the dream grows must remain unconscious, like backstage machinery in the theater, if a strong message is to emerge.

This was a most useful lesson, and now I always advise students to be satisfied when they have obtained a useful message from the dream—the fruit of the tree—which moves them to take constructive action in their lives. This does not mean being slipshod in dream interpretation—the major symbols and events of a dream must make sense for it to be understood—but there is no need to push the river in trying to make every single dream item meaningful.

DREAM POWER AS *SPIRITUS RECTOR*

The last example shows dream power moving the conscious mind to get a better sense of proportion, and this is a very important function of dreams. Jung was so impressed by it that he often used to say that the unconscious at times acts like the *spiritus rectòr* of the medieval mystical theologians—an inner wisdom that serves as a regulating or balancing force to counteract one-sided or extreme attitudes of the conscious mind.

I always warn people of this when they tell me they intend to program their dreams to cheer themselves up if they feel low. This kind of request or suggestion to dream power will work only as a temporary measure to give you the encouragement and energy you need for coping with a problem, but it cannot be used as a permanent escape from reality. Dream power will override a request under these circumstances and throw up instead a dream showing what your real problem is —and it often does it with a wry sense of humor. When John asked for a flying dream, he did indeed fly all night—by Pan Am, British Airways, and every other line imaginable! This was not a joke, and John got the message, namely that for most of his life he had relied on the power of industry and technology to get him "high" in the world and was not yet accustomed to the thought of being able to "fly" under his own power. And when I once asked dream power to transport me to a good place in my dreams, I spent all night literally being transported there by car—but I never arrived. The dream was telling me that if I really wanted to get to good places in my life, I should slow down, stop rushing around, and become more centered. As always, the moral is "as in life, so in dreams."

So always use dream power to get in touch with your heart's wisdom for moving forward in life, but don't expect it to offer you an escape or to act like a genie who does everything for you by magic. I had a sharp lesson in this respect years ago when I seemed to be experiencing something akin to the dark night of the soul in which life seemed to lack all sense of purpose. I asked dream power for light—and I got it, together with a message that blew my mind. I dreamed that my mother left me alone in the house and the lights went out. I fumbled around in the closet for a while, then became frightened by the darkness, and decided to go out until she returned. Later I saw a light under the door and knew that in her usual efficient manner, she had fixed the lights.

The dream made me realize that all my life I had been dependent for "light" on my mother, becoming ill and depressed whenever I felt I was not living up to her high ideals drilled into me as a child. If I ever lost my temper, her voice within me accused me of being awful, incapable of improvement, and possibly suffering from some glandular disease; whenever I relaxed, the voice would accuse me of laziness and so on, giving me the feeling that the "light" of God's approval had gone out of my life. My "dark night" was, in fact, nothing more than self-punishment for my sins, and light would be forthcoming when "mother" consented to forgive me. The dream showed me that it was quite unnecessary to continue suffering in this passive way, and that I could turn my own light on whenever I wanted. I now know that whenever darkness seems to overshadow my life, the solution is not to ask dream power or God to remove it, but to look inside myself and discover what I have done to evoke the reproaches of my mother's voice within—and then turn on the light either by seeing through her impossible demands, or by forgiving myself again and again and again and again . . . for there comes a time when you have to take your life in your arms.

DREAM POWER AS FAMILY COUNSELOR

The majority of the heart's concerns are with personal relationships, and this is an area of life where requests to dream power can be immensely helpful. The technique is essentially the same as that described at the beginning of the chapter: relax in bed and ask dream power to "Please send me a dream tonight to show how I feel about my marriage . . . my husband . . . my girlfriend . . ." or whatever. Bear in mind that the request must be a serious one involving your heart.

Rob, who participated in one of our student groups, came to class one morning saying that he had tried our request experiment and it had failed. Having just emerged from a very traumatic love affair, Rob had embarked on a new relationship with a girl student, and not wishing to repeat the last bad experience, he had asked dream power for his heart's thoughts on the subject. He dreamed of making bread, a thing he did often in waking life and enjoyed. We asked him to take one of two chairs facing each other, put the "bread" in the opposite chair and talk to it. Moving his hands and arms as though kneading dough, he began, "I'm making you and kneading you, and I'm enjoying kneading you . . . ," stopping abruptly as the class broke into laughter at dream power's splendid pun. Rob resonated totally with the feeling that he enjoyed needing (kneading) the girl, and moving to the opposite chair and giving the "bread" a voice, he continued, "And I enjoy being kneaded by you," a happy start to a hopefully happy relationship.

Joan, a member of a weekend workshop, asked dream power to tell her how she felt in her heart about her marriage. Her head knew that it was unsatisfactory, but she had not contemplated divorce on account of the children. She dreamed that the famous Chicago store, Marshall Field's, was on fire and that children were being burned to death inside. She got them out, but as she left the store, she had a vision of a woman sitting in a small, dark room higher up in the building doing accounts, quite oblivious to the danger.

As she had always thought of Marshall Field's as a "family" store, she had no difficulty in relating the dream to her marriage. While her head recognized it as unsatisfactory, her heart saw it as a disaster—"up in flames," with the children being burned to death within it. Joan thought the woman doing the accounts upstairs was the part of herself that did not want to face the situation honestly—the housekeeper who continued to do the accounts while the home burned down around her. Perhaps the major revelation to Joan was the way her heart saw the children being destroyed by the marriage, while her head had been thinking they would be destroyed if she left it. I don't know the outcome. At the time it was enough that Joan should understand her heart's thoughts on the issue, and having got herself together, then make some rational decision.

DREAM POWER AS VOCATIONAL COUNSELOR

Another area in which requests to dream power can be very useful is that of making decisions about the future course of your life—for example, if you are contemplating a new job or change of life-style. Here again, it is essential to remember that dream power is not some kind of oracle that will absolve the head from responsibility for rational decision. It simply helps you get yourself together by making you aware of your heart's thoughts—and your heart may have quite different standards of judgment from your head. Wilson Van Dusen puts this very sensibly in his book *The Natural Depth in Man:*

> In general, the inner is concerned with the whole quality and style of your life. It is not very involved in whether you take this job or that job, etc., unless the jobs feel very different and represent fundamentally different approaches to life. The inner will talk about directions you are now taking or contemplating, but it is rarely predictive. If you want it to comment on what stocks to buy, it is more likely to kid you about the foolishness of your aims. Though it may know the future, it rarely shows it. It is more like a psychotherapist or a very good friend. Its fundamental concern is with the design and quality of your life, not in making you rich or famous. And it is in the design and quality of your life that it is a master consultant. . . .

When Janet and Dan attended a weekend workshop, they had just such a problem on their minds. Dan had been offered a new job in Ohio with the Republican party, and they were both of two minds about taking it. To Janet, in particular, it would mean the uprooting of a home and a complete change of life-style, so they considered it quite legitimate to ask dream power for help. The following morning Janet reported a dream in which she was in a gambling den playing poker. When someone called her hand, she replied, "I'm not the Cincinnati kid!," revealing her heart's thought that Ohio was not the place for her. And when Dan came to the part in his own long dream where he found himself with a group of people all lifting their arms in the Nixon salute while he remained paralyzed, the whole group, including Janet and Dan, considered their problem solved. When we said good-bye, both were decided not to take the new job.

Very often the issue is not so much the actual choice of job, but the direction one's life is taking or should take. Paul, the psychiatrist in charge of Janet and Dan's weekend, asked dream power how it felt about his life. He dreamed he was walking down the Via Condotti in Rome, where he had just spent a holiday. The streets were straight and narrow with lots of traffic, and he particularly noted the clear line of demarcation between the vivid blue of the sky and the red earthiness of the street. He interpreted the dream as his heart's thought that his life at the moment should remain with his patients—for the Italian word "con" means "with," and his patients could clearly be called "dotty." He associated the clear blue sky with the world of the spirit which he would sometime like to explore, and the red streets with the "straight and narrow," down-to-earth job he felt he should complete with his "dotty" patients in the busy world of human conflict.

One of our students, Jefferson, asked dream power to tell him why he was at college and the consequences of staying there. He dreamed:

> I am being given electric shock therapy here at college. My parents and everyone involved seem sincerely to have my interest at heart and feel it is the only cure for a condition I hadn't realized I had. I am given two treatments and remember them being extremely painful. I realize that I am not the zombie I am supposed to be, but fake it in hopes of escape . . . then I telephone my mother trying to convince her that shock treatment is too severe and I am sane anyway. I remember feeling in the dream that the call was a futile effort.

Jeff wrote in his end-of-term report:

> Before I came to this college, I had a vision of people in the process of transcending the petty competitions and aggressive attitudes that I had experienced in high school and a year of traditional college. I imagined we would all be sensitive and aware of some higher truth, whatever it could be. But the students here all seem emotionally closed and detached, like zombies. I slip into their pattern because everyone, including my parents, believes it good for me. But I know that they are sick and I am sane . . . but it's useless trying to convince them. They say I must prepare myself for the big competitive world, but I don't want to join it. What I'm really interested in is sculpture. . . .

I don't know what decision Jeff took, nor just what it was about college that made him feel so strongly, but he would obviously pay a terrible price if his parents or anyone else tried to persuade him to carry on simply because they considered his feelings absurd. Emotions cannot be changed by an act of will just because people think they are mistaken, and feelings as strong as these indicate that for Jeff at least the college was proving horribly disharmonious and painful, no matter how many other students felt differently. When our hearts throw up such violent distress signals, a radical change of life-style is usually indicated—and for young people today alternatives are possible.

Ann, a professor of psychology who attended several of our workshops and is now using dream power in her own classes, had a more specific request. I quote verbatim from her letter:

Request: Dream power, I have a terrible headache. One of my classes is going terribly because of two people hostile to what I'm teaching. I'm superanxious about *something!* What is happening? One student accused me of being manipulative and crusading only for my own approach to psychology. Is this true, dream power?

Dream: I am the first woman astronaut going to the moon. I land my spacecraft and proceed to dash back and forth from the spacecraft to the surface with presents. (I'm giving the moon its first Christmas tree and all.) I'm so excited about all my gift giving that I realize I have forgotten to maintain radio contact with earth—and I have forgotten to perform the scientific experiments.

I go wandering and exploring and come across a little blue house with a picket fence. As I approach the door a "menopausal woman" yells and shouts at me and her husband inside looks at me with a growl. I've done nothing to upset them. They just seem like angry people. I tell the woman I'm the first woman astronaut going to the moon and she seems unimpressed except with the fact that I'm the first *woman* astronaut. I ask to see a newspaper and reading it, find I'm not on the moon but rather am in Sweden. I say, "Oh, well, that's O.K. I've never been here before."

Interpretation: O.K., Ann, you're doing lots of "firsts" in your approach to teaching psychology at this college. You are special. But in this class, you've lost radio contact with earth because you were so excited by all your handing out of gifts and new experiences to your students. You need to stay in better touch with the earthy side of things—how your students are feeling about the course, whether you've gone beyond them in your overenthusiasm. And you need to be fairer to the scientific part of psychology—more objectivity in representing concepts you don't agree with. The man and woman students giving you the trouble are on their own anger trips. She's menopausal, just has to be that way. This has never happened to you before, you've never been to a place (a class of yours) so characterized by the anarchy you associate with Sweden, so it's O.K. Just stay in radio contact next time.

It's great to feel I'm the first woman astronaut. I haven't got to the moon yet, but I'm going. . . .

DREAM POWER AS SPIRITUAL COUNSELOR

When we were in California we were delighted to meet and make friends with an unorthodox young psychiatrist, Dr. Douglas M. Gregg, who runs the San Diego Medical Hypnosis Center. Dr. Gregg's major interest is in the coupled use of dreams and hypnosis, but he also runs encounter, marital, and meditation groups, as well as studying astrology and related subjects with his wife, MacGlenna. I was intrigued to discover that he had been using the dream request technique for years in his own life and in his practice with patients. When a new patient comes for treatment, he is taught to relax and enter a light state of hypnosis, and then Dr. Gregg suggests to him that he will dream about the basic cause of his problem.

In his booklet *Hypnosis, Dreams and Dream Interpretation* Dr. Gregg explains how this can be done by the patient himself at home:

> Either by waking suggestion or post-hypnotic suggestion anyone can program his subconscious mind to bring forth dreams that are vivid enough to either awaken him or to remain in his consciousness upon awakening in the morning. In addition to being suggested or requested, dreams can also be given direction. This process is extremely valuable for defining and solving problems rapidly and completely by tracing problem areas

> to their respective causes. With the proper use of dreams it is possible to structure a program to rebuild the patient by removing the cause of certain physical and emotional problems rather than simply attempting to relieve symptoms.

Reading this, I was reminded of Wilhelm Stekel's observation, made in his two-volume work *The Interpretation of Dreams,* that every reasonably vivid dream gives a picture of the patient's parapathy (neurosis), and so I decided to use Dr. Gregg's technique to ask dream power about my own major life conflict. I dreamed I was in a restaurant getting absolutely no service. I became very angry and decided to go elsewhere, but first I thought I would write a note expressing my dissatisfaction and leave it on the table. With great difficulty I wrote a word across a piece of paper, and when I had finished saw the word "victim" in large letters scrawled over the page.

I was able to resonate to the obvious message—namely that I tended to seize on any small inconveniences of life to play the game of "Victim." Instead of seeing them as ordinary life incidents, largely accidental in nature, I would take them as deliberate rejections of my essential worth. My major neurosis was obviously a paranoid tendency, and it made sense to me when I looked back over many events in my life.

In the case of every neurosis we have to ask the question "What is it doing for the patient?" In other words, the patient must be getting *something* out of it for him to hang onto it so desperately. (Psychoanalysts call this "secondary gain.") We know that illness elicits sympathy, care, and attention from those around us and also absolves us from all kinds of unpleasant chores. So I asked myself what the game of victim could be doing for me and concluded that in some masochistic way I must actually enjoy suffering, though I could not really understand why.

I did not return to the subject until I listened to a taped

lecture by Elsie Sechrist, author of *Dreams—Your Magic Mirror,* in which she suggested that we should ask our dreams to show us the greatest stumbling block to our spiritual growth. While I would have been happier to ask about my most positive aspect, I decided to try her experiment and add my own at a later date—and I was very glad indeed that I took her advice, for it enabled me to solve the unfinished "victim" mystery. I had three dreams the same night in response to my request:

> *Dream 1:* Elsie Sechrist is standing outside my window peering in accusingly.

> *Dream 2:* Dr. Calvin Hall (the dream psychologist) has lent me his house in Santa Cruz for the night while he stays at his mother's place. I sleep late in the morning, and rush round frantically looking for the happy badge he had given me the previous night, over which he had scribbled the number 13.

> *Dream 3:* I sweep out my closet and find some dead plants we call Crown of Thorns. I burn them and feel really good about it.

Working on the dreams, I recalled how some of Edgar Cayce's followers interpret looking in through a window as an overview of the subconscious or inner life—so here was Elsie Sechrist, whose dream work derives from Cayce, getting a good look at my subconscious. Using the creative monologue technique, I asked her what she was doing, and "she" replied, "Dream power sent me along to have a look at Ann's subconscious mind because she wanted to know her greatest stumbling block to spiritual growth. I'm very good at putting my finger on weaknesses, faults, and sins . . . and I must say there are plenty here. In fact, there are so many that I'm going to find it difficult pinpointing the major one!" As that was as far as I could get with this dream, I moved on to the next one to discover "Elsie's" choice of subconscious weakness.

As I hadn't seen Dr. Hall for over a year, the dream could not refer to him personally, nor had he given me a happy badge or the use of his house. My immediate association was to his name, Calvin, which could stand for Puritanism, a thought that was at once corroborated by the fact that Santa Cruz is Spanish for "holy cross," a symbol of suffering. So the dream seemed to link up with the one I had when I first tried Dr. Gregg's experiment—the main obstacle to my spiritual growth was my neurotic desire to suffer. I mentally apologized to poor Dr. Hall for landing him with this association, since in real life he would be the last person to try to cancel happiness with bad luck, which was how I saw his writing the number 13 on a happy badge.

I was very grateful to be given the clue that my "masochism" had something to do with a puritanical belief that happiness should be superseded by suffering—and I was delighted to know that I had within me the resources to make a "clean sweep" and give the Crown of Thorns—another symbol of suffering—the treatment it should have had years ago. But I was still unclear as to what secondary gain I could be getting from my suffering, and so I asked dream power for clarification.

The answer came loud and clear in a dream in which I was a barmaid in a miserable restaurant serving drinks and washing up for about fifty cents an hour. On waking, I immediately associated serving drinks with "serving the Spirit" in a religious sense and realized how all my life I must have labored under the (unconscious) impression that I could truly serve God only by being unhappy, unsuccessful, and preferably hungry—for had I not been taught long ago that God loves the poor, the hungry, the mourners, the persecuted, the reviled, and the rejected? These teachings and many others like them had penetrated my child's mind and had caused me all my life to write 13 across my happy badges, as it is well known that God does not love the happy and successful people of this world.

Thanks to dream power, I have now swept this crown of thorns out of the closet of my unconscious and am now able to enjoy life without feeling sinful. But it is sad that I felt obliged to play the victim game to get God's love in the first place.

My next dream saw me back at the restaurant washing the dishes, but this time it dawned on me in the dream that I was a psychologist and author and didn't have to do this kind of work any more—so I gave notice! The dream was what Dr. Gregg would call a "verification dream," one that occurs after a problem is solved and confirms that the solution is correct. If you are ever in doubt about an interpretation, you can always ask for such a verification.

DREAM POWER AND CREATIVE INSPIRATION

The idea that dreams can provide creative inspiration for works of art, inventions, and scientific discoveries is age-old and there are a number of famous instances that are often quoted: Robert Louis Stevenson, who dreamed the plot of *Dr. Jekyll and Mr. Hyde;* Elias Howe, who dreamed the invention of the sewing machine; Kekule, who grasped the ring formation of atoms in benzene after dreaming of a snake swallowing its own tail; and so on. I have come across a number of other cases of this kind in the course of my work: Margaret Mead told us that she kept a dream diary for years and often received guidance from her dreams; Meredith Sabini, a doctoral student in California, described in her 1972 dissertation on dream groups how a dream led her to change her thesis subject and how she received hints and help from her dreams while writing it; and Kreskin, the famous mentalist, told me that many of his "effects" come to him in dreams.

A rather more unusual kind of inspiration came to my husband while we were staying with John Benson Brooks, the composer, and his wife, Peggy, in New York. He had a very strange dream in which he was instructed to inform John Benson that his next song should be titled "Love Becomes an Owl." John Benson took the dream very seriously indeed, with the result that there is now a beautiful song, complete with words and music, of that name. We call it our "Dream Song."

Whether the dreaming mind itself makes these creative acts or whether they are produced at the back of the mind during the day and then thrown up in dreams is an open question at the moment. One of the most frequently cited tales of dream inspiration in science is that of Otto Loewi, the Nobel Prize-winning biologist who was supposedly led to the idea of chemical transmission of nerve impulses in a dream—but most of those who retail this story are evidently unaware of the fact that he later realized he had consciously worked out this idea eighteen years earlier and then completely forgotten about it. In this case the dream was obviously throwing up an idea already formed in waking life, but was nonetheless useful for that. (Many other fascinating cases of cryptomnesia—hidden memory—are quoted by Robert Merton in his book *The Sociology of Science.*)

Those who receive creative inspiration in dreams are almost always deeply preoccupied with their problem in waking life, so it should not come as a surprise to know that we can ask our dreams for help in this area. Sometimes the result is not a direct inspiration, but a revelation of what is blocking our creative energy in waking life—but more often than not, dream power obliges with a solution. For example, Barbara Seaman, editor of a nationwide journal, was worried when I met her at a literary gathering in New York because she could not find the "right" title for her new book, and I quote the story in her own words:

> I had been trying for six months to find a title and indeed finding a title was a subject of some discussion at the Society of Magazine Writers meeting we all attended. Ann advised me to think, as I was falling asleep, that I would find a title in my dreams. She said that when I woke up, I probably wouldn't find that the title had appeared like skywriting or anything like that. But rather, I might retain sharp images from my dreams and that I should associate to them. The dream I remembered the next day was of Diana or some other Greek goddess running through the woods. This was a most unusual dream for me as most of the dreams I recall are ones in which I am the central character. Diana, or whoever, was wearing the traditional flowing garb through which you could see both her shapely female body and her strong but lovely leg and arm muscles. She was very feminine but also very strong and happy. I associated to her, as Ann had instructed, and the first thing I thought of was feminine and free. Then I thought feminine sounded too cutesy and not sufficiently strong. And so my next thought was "free and female."

Free and Female, which was published soon afterward, has been a great success and is now out in paperback. In the preface, Barbara gives me acknowledgment for help with the title—but thanks should really go to dream power. Evidently her dreaming mind was able to find the exact motif she needed when her waking mind was completely stumped—she had three hundred titles written down when we met and not one of them was "right." Barbara was so impressed by the power of dreams that she has used them ever since to solve major conflicts in her life.

Dreams have always played a major role in artistic creativity and have inspired such artists as Leo Katz and Irene

Rice Pereira to produce some of their finest works. When I met him, Leo Katz told me that many of his paintings are in fact almost exact replicas of dreams or visions experienced in altered states of consciousness. One of his most famous paintings, *Metamorphosis 1942,* was produced after a very wonderful and vivid dream which made such an impression on him that he was able to remember every detail. The dream now appears on a 100-square-foot canvas and is considered by many to be the great prophetic painting of the twentieth century.

The life of Irene Rice Pereira, poet and artist who died in 1970, was also inseparable from her work, and so it is not surprising that her dreams, too, were a constant source of inspiration and guidance. Throughout her life she recorded her dreams in a diary, and I hope to have more information about these when her papers are made accessible to researchers. Before falling asleep at night she would assign tasks to her dreaming mind and would arise before dawn to constellate the ideas and impressions that emerged. In her illustrated portfolio *The Lapis** she describes a dream that changed her life by bringing about a creative breakthrough in artistic technique:

> I crossed a bridge. The water was surging on both sides of the bridge. This made it very dangerous. The bridge was very close to the surging water; but it was a sturdy bridge so I crossed it safely.
>
> Now I am in a place with many people. Someone is telling me about a dream and suggests that I paint it. I say, "I cannot paint someone else's dream, I can only paint what is really mine." Now I see the suggested picture. It is an image, quite beautiful, but it is not real because the image is lying flat.
>
> Now, as I contemplate the scene, I see the whole thing. The

*Washington, D.C: The Corcoran Gallery of Art, 1957.

image is lying flat on a circular island, in the center of a most beautiful circular blue lake, surrounded by a verdant landscape of trees and an ethereal sky. It is beauty. It has a mystical quality of atmosphere in a spherical depth. But I am distressed because the image is lying flat.

Now, as I look and contemplate this scene, the image of its own motion is lifted into a vertical position. I see it now. It is an oval stone, life size, smooth and polished. It is about one foot in thickness, and 6, 8 or 10 feet tall. The stone monument is polished Lapis Lazuli with a white figure carved or incised into it. It was really so very beautiful. The figure looked like the cameo carving in the Lapis Lazuli of a ring, only life size. The figure incised in the stone looked Archaic Greek, or like an ancient prophet. What disappointed me was the fact that the reverse side of the Lapis Lazuli stone monument was unknown, i.e. it had no figure on it.

Now, it is all clear to me. I see the picture of what I am supposed to paint; it comes into view. It is beautifully illuminated and alive and set in a spherical depth of illuminated ethereal atmosphere.

On waking and meditating, and then starting to work, she realized she had been brought up against a fundamental problem of artistic imagination, for she found that she was unable to transfer the wholeness of her dream vision to paper without destroying parts of it. The trees in the foreground surrounding the lake had to be removed for the stone to be seen at all, and part of the background scenery was necessarily blocked by the stone itself. She spent many weeks trying to arrive at some kind of solution, studying all kinds of scientific works on perspective and investigating the relation between the history of painting and mankind's changing philosophies of nature. She concluded that geometric systems of thought have avoided the infinite, whereas visual and intuitive perception involves the infinite in enabling us to see the Whole. Eventually, the dream image drove her to a whole new concept of using light and space in her work and to a new understanding of her philosophy as an artist.

Robert Masters and Jean Houston at the Foundation for Mind Research in New York have carried out many fascinating experiments on the use of hypnotic and kindred techniques for overcoming creativity blocks in artists, writers, and musicians, and one especially interesting discovery they have made is that when a subject is given the suggestion that time is running slowly, he will complete a creative act in a fraction of the clock time normally needed to do so. This must involve the speeding up of brain processes, a hypothesis put forward in their books *The Varieties of Psychedelic Experience* and *Mind Games.* If I am correct that REM dreaming involves a similar speeding up of brain processes, this could mean that dreams offer immense potentialities for help to those engaged in creative work.

Masters and Houston believe that major blocks to creativity are caused by the mind's energies becoming locked in some deep inner conflict rooted in what they call the "cultural trance," a fixation on outmoded values and institutions. This links directly with my own experience that when dreams are used to confront and overcome such inner conflicts, an immense amount of creative energy is released. Ernest Becker, the philosopher, in a remarkable deathbed interview with Sam Keen in *Psychology Today* in April 1974, pinpointed this phenomenon in a very interesting way: "In my mid-thirties," he said, "I suddenly started to experience great anxiety, and I wanted to find out why. So I took a pad and pencil to bed and when I would wake up in the middle of the night with a really striking dream I would write it down, and write out what feeling I had at certain points in the dream. Gradually my dream messages, my unconscious, told me what was bothering me—I was living by delegated powers. My power sources were not my own and they were, in effect, defunct." In the next part of the book, I shall be showing what some of these delegated powers are—those inner voices of parents, teachers, churches, and other authorities long since defunct and irrelevant to the individual's present life and growth—and how dreams can help us break out of the cultural trance to reclaim our lost independence, vitality, and creativity.

PART III

GAMES FOR ADVANCED PLAYERS

«9»

THE HOUNDS OF HELL

It is not the lie that passeth through the mind, but the lie that sinketh in and settleth in it, that doth the hurt.

—Francis Bacon

Some time ago, our local newspaper carried an episode of Jack Moore's syndicated cartoon strip *Kelly* which touched on a major psychological problem. It began with Kelly, the small boy, asking Duke, the big puppy, why he is dressed in clothes.

"Ah heard this voice sayin' Duuuuuke, Duuuuuke . . . ," says the puppy gesturing toward the heavens.

"And what did *you* say?" asks Kelly.

"Whaaaaaat, whaaaaaat?" replies Duke, with his hand over his ear as though listening. "Then th' voice told me t' go out in th' world an' cover up all filthy an' disgusting things."

"What things?" asks Kelly.

"Th' voice," says Duke, "said Ah could start by coverin' up mah BODY!"

"What did the voice sound like?" asks Kelly.

"It was kind of a high-pitched hysterical shrill," replies Duke.

"Mom?" asks Kelly.

"Mom," replies Duke.

One of the most important lessons we can learn from dreams is how our lives are ruled, our energies wasted, and our happiness spoiled by voices from the past whose conditioning is totally inappropriate to the present. Like Duke, most of us go around thinking the things we believe in and do are decreed by God, or by some inescapable law of right behavior embedded in our bones, when we are merely acting out injunctions of parents, teachers, religious gurus, or other authorities of the past, so deeply ingrained in our minds that we have come to take them for granted. All psychotherapy is concerned in one way or another with the debilitating conflicts that go on inside people when these ingrained authority voices cut across really basic needs of the personality, but it is not only "sick" or "neurotic" people who suffer in this way—we are all restricted much of the time in present-day society. In fact, the health or sickness of any society can probably be judged by the extent to which the lives of its members are bedeviled by such life-wasting inner warfare—and I suspect this may be the central psychological problem of the whole human species.

Freud came upon this problem as soon as he started investigating the unconscious, and his model of man shows a raging, instinctual id of sexual and aggressive urges constantly striving for expression against the moral dictates of a culturally conditioned superego. The only solution he saw was the sublimation of these basic needs into culturally creative activity like art or into harmless aggressive activities like sport. This is a very pessimistic view of human nature, and one that offers no options for transcendence. In my view, the psychologist who expressed the problem most clearly and simply, yet also investigated it most deeply and came up with the possibility of transcending it, was Fritz Perls, the "finder" of Gestalt therapy. He called the internal authority voices the "topdogs" of the mind, trying continually yet fruitlessly to impose their will on the rest of the personality, which then behaves like an "underdog" wanting to keep topdog's approval and at the same time trying to get its own way. I prefer this to Freud's model of a "superego" struggling with instinct, and even to Eric Berne's idea of an inner "parent recording" constantly upbraiding an inner "child" for unacceptable behavior. Perls's model brings out the fact that topdogs can be of many different kinds and that each of us will have to deal with several quite separate ones during the course of our lives—not just the very basic taboos implanted by parents about sex and emotional control, but also the injunctions and values of school friends, neighbors, compatriots, religious teachers, and so on. In this and later chapters, I shall be describing some of the many and varied breeds of topdog that have emerged, along with their corresponding underdogs, in our own dream workshops.

Perls found that topdogs and underdogs often emerge spontaneously when the dreamer enacts his dream in waking fantasy by means of the dialogue technique, in which the dreamer gives a voice to two of his dream characters and lets them converse with each other. While dream dialogue itself was no new discovery on Perls's part, his method of focusing on these inner conflicts and working through them to a creative resolution was a major breakthrough in dream psychology, certainly giving our own dream work a deeper dimension. In Perls's words, it gets you to the center of the personality in a very short time—though working through to freedom may take much longer!

The following example from one of our workshops shows such a dialogue in action, bringing to light one of the commonest breeds of topdog in our Western civilization. I have deliberately chosen a short, simple dream which resulted in a typical topdog/underdog conflict in order to illustrate my points.

TOPDOG IN THE MANGER

Sam, a middle-aged businessman, dreamed he was fishing off his local pier when a black car on the road behind him started honking. At first he tried to ignore it, but soon became irritated by its persistence and went home.

The black car reminded Sam of the one used by the president of his company, but he could not resonate to the notion that either the president or his car was on his mind at the time of the dream, which occurred during a weekend. The dream seemed rather to express a conflict in his own heart between duty to work and his love of fishing as a relaxation, and he confessed that it had probably been sparked off by a recent secret plan to get away alone for a long weekend of fishing, escaping from both work and family burdens. This seemed to be only the beginning of the story, however, so we asked Sam to sit in one chair and place the president's car in the chair opposite and talk to it. The dialogue went something like this:

Sam: Why do you keep honking like that? You've ruined my fishing.

Car: (played by Sam, moving to the other chair) And I should think so too. You shouldn't be out fishing with all that work to be done at home.

Sam: (moving back to original chair) But I need a break. I've really been overworking lately . . .

Car: Nonsense. Only weaklings need breaks. Come on, pull yourself together.

Sam: (petulantly) But I don't want to work today. I want to fish. I was really enjoying myself.

Car: You're not in this world to enjoy yourself. And you have next week's contracts to prepare.

Sam: (pleading) Yes, I know, but I was going to do those

tomorrow. Look, just shut up for now and let me fish in peace today, and I promise to catch up on work tomorrow.

Car: Certainly not. If you take time off today, it will become a habit. And always remember—"never put off till tomorrow what you can do today." And remember what Baba Ram Dass says in that book you've just been reading—you must blow out sloth to grow in spiritual stature.

Sam: (whining) Yes, but . . . well, you know—"all work and no play makes Jack a dull boy."

Car: (sternly) That's just the point, Sam. When are you going to realize that you aren't a boy any longer? You're a grown man with responsibilities to your wife and family. You can't live for yourself alone. Your poor wife will die of disappointment if you don't get that promotion. She'll lose all respect for you . . . (sighs) . . . I wish you would understand that I'm only trying to help you because I know what's good for you.

Sam: Oh dear, yes, I know . . . it means a lot to Ellie.

Car: And your poor father would turn in his grave if he thought you were turning into a slob—after all he did for you.

Sam: Oh, yes, I know . . . oh dear.

At some point in every dialogue like this the true identity of the topdogging voice behind the mask of the bullying dream image suddenly becomes clear. In this case the car, which at one level could be understood as Sam's "drive" for success in his business, turned out to be nagging him with the injunctions instilled into him long ago by his father, a successful businessman who had done his best to mold Sam in his image. As Sam had been rewarded with love and approval whenever he lived up to his father's standards, and deprived of them whenever he protested or rebelled, he had been forced throughout childhood and youth to choose between the satisfaction of his own basic easygoing nature and his father's approval. As withdrawal of parental love is akin to psychological death in a child, Sam chose his father's approval, entered the business, and attempted to live a replica of his father's life. And although the real father was long since dead of a heart attack brought on by overwork, his voice had lived on within Sam to nag him whenever he deviated from the straight and narrow path of the puritan work ethic. Of course, Sam tried from time to time to justify himself in taking a little relaxation on the ground that he needed it for health and recreation, but invariably the nagging voice would come in very quickly like a dog in the manger to spoil his enjoyment.

Before moving on to show how such a topdog/underdog conflict is resolved, it is worth noting the basic characteristics of topdog and underdog as shown in this particular dialogue (which is typical), so that whenever you meet characters like them in your own dream work, you will recognize the pattern even though your problem may be about something quite different.

TOPDOG

1. His favorite words are "should" and "ought"—"you *shouldn't* be out fishing, you *ought* to be working."

2. He always knows what is best for you and insists that what he is doing is for your own good.

3. He adores clichés, platitudes, and quotations and will use as many as possible to corroborate his case—"never put off till tomorrow what you can do today." He also has a really nasty habit of quoting the people you most admire—"Baba Ram Dass says . . ."

4. He preaches perfection, nothing less.

5. His whole case rests on a *catastrophic expectation* that if

you give in to your need, it will rapidly overwhelm and destroy you—"You will become a slob"—or "you will lose the respect of your fellows"—or "your wife will lose all respect for you."

6. He is judgmental, making dogmatic statements like "you are irresponsible" and "you are a lazy slob," instead of expressing his own *personal* opinion by saying, "I don't like to see you relaxing, it worries me."

7. He is very fond of little sermons—"you are not in this world to enjoy yourself" or "you can't live for yourself alone." If underdog shows any sign of trying to stand up to him or rebel against him, he will often adopt a look of hurt reproach and say, "How can you be so cruel as to treat me like this when I have only your best interests at heart?"—a trick used by real parents and others who are determined to maintain control when straight bullying fails.

In summary, topdog is a preacher, a dogmatist, a perfectionist, and a bully—but if you watch him closely, he is also a hypnotist and an exaggerating hypocrite. He is everything Jesus condemned in the Pharisees.

UNDERDOG

1. He is overwhelmed by topdog's case, hence is apologetic and makes excuses—"but I've been overworking lately."

2. He also knows how strong his needs are, however, so he whines and pleads—"let me fish today . . ."

3. He tends to think that topdog has both right and might on his side, so resorts to promises—"I promise to catch up on work tomorrow."

4. When pushed, he fights back unconvincingly, often trying to answer topdog's clichés with opposite ones—"all work and no play makes Jack a dull boy."

5. He tries to win topdog's forgiveness by self-castigation—

"oh dear, yes, I know I'm a slob . . ."

6. He usually ends by giving in to topdog, without whose approval he feels he cannot live—but . . .

7. . . . if you look and listen carefully, you will detect that he hasn't really given in completely—at the back of his head is the hope of finding some way of getting what he wants without incurring topdog's disapproval.

In summary, underdog (beautifully personified by Jack Moore's Duke) is the archetypal coward, victim, masochist, and dupe—but he is also a cheat who will resort to underhand, devious means to get what he wants despite his overt compliance.

This last point is crucial to understanding why human life is so continually beset by these conflicts and why Perls called the fight between topdog and underdog a "self-torture game" played out endlessly by these two clowns of the personality—a game of "should I, shouldn't I"; "can I, can't I"; "will I, won't I." In dialogues like the one I have described, a careful observer can usually detect underdog's secret determination to find a way of getting what he wants, by noticing how the person is sitting when he is in the underdog chair or the expression on his face, which often belies his humble and contrite words—and if you are doing the dialogue on your own, you may be able to catch the sly thought at the back of your head. A cartoonist would portray the situation by putting a balloon over underdog's head containing the words: "*Thinks:* if I really work myself silly I may get to a state where the doctor orders me to take a fishing holiday" or whatever.

For invariably in life underdog *does* find a way to get what he is after by careful manipulation of circumstances, so that what he wants happens without his having to take responsibility for it. A great many illnesses can be traced to this process, usually when underdog wants to rest or be looked after. But that is only one of his many underhand devices. Another is somehow to contrive that his partner or family call upon him to do what he wants to do, so he can say in effect to topdog, "Of course, I would never dream of doing this for myself, but . . ." If his needs are sexual, he contrives to get himself seduced, or somehow arranges to find himself in circumstances of "overwhelming temptation," promising never to give way to such weakness again. Our manipulation of ourselves in this way is usually justified and dignified by the word "conscience"—which never prevented anyone from doing what he wanted, merely from enjoying it!

This pattern of constant failure to heed the voice of "conscience" and live up to our inner ideals of perfection has been obvious all down the ages, and St. Paul gave it classical expression when he said, "The good that I would, I do not, and the evil that I would not, that I do." But this traditional way of putting it really misrepresents the whole situation, which is why mankind has made such little progress in this area.

The vital point which is obscured by the traditional phraseology is that an impasse arises *only* when the good we are trying to do is in some way unrealistic, either for the individual or for society as a whole, like asking an almond tree to produce oranges. The villain of the piece is topdog, who preaches his own prejudice in favor of oranges, as though it were part of the divine (or natural) law of growth for all plants. Parents, teachers, religious gurus, and other authority figures constantly do this in reality, but the impression is reinforced in the mind of the growing child by the fact that he is totally dependent on the approval of adult figures for his survival and has no way of knowing that their injunctions are not indeed divine. Sometimes topdog makes demands so completely at odds with the individual's real nature—like asking a naturally introverted child to become extroverted, or a low-energy type to become high-powered and driving, or an artistically inclined person to be an intellectual, or a natural yogi to be a commissar, and vice versa—that the result is some kind of breakdown.

But normally topdog's injunctions really do appeal to something in the individual's makeup, which is why poor underdog finds it so hard to go against him. The error lies in the belief that topdog's way of achieving this good—by more and more controls and discipline—is correct, instead of realizing that it is the controls themselves which are stultifying and hinder growth and progress. If an inner voice whispers gently that you could perhaps make more effort to achieve some desired goal and you are able to do this without too much strain, then no topdog/underdog impasse arises, which means that the goal is realistic *for you,* though it may not be for anyone else. But if you constantly fail to live up to your inner ideal, then you are doing something wrong, and it is time to sidestep and change your ideal instead of forcing yourself into a Procrustean bed made for someone else. We grow by expanding, not diminishing ourselves.

The great mistake of all moralizers, from St. Paul to the average parent or teacher, is to treat underdog's resistance to topdog as an evil to be overcome, on the model of a wild horse that refuses to be tamed, when in reality it is like the almond's "refusal" to become an orange. The moralizer (and the internal topdog inside each of us) sees only the misshapen and ungainly growths that result from the conflict between his efforts to train the plant and its own inherent nature, and he never stops to think that the plant has the capacity to grow into something quite beautiful on its own if only he were prepared to help it instead of imposing his stereotype upon it. Instead, he concludes that the need is for still more rigorous training in the orange-ward direction, which simply makes matters worse. The really great religious thinkers have always recognized this—and modern psychology merely gives detailed chapter and verse to the wisdom of Lao-tse when he wrote:

If I keep from meddling with people, they will take care of themselves,
If I keep from commanding people, they behave themselves,
If I keep from preaching at people, they improve themselves,
If I keep from imposing on people, they become themselves.

—or to the more modern teaching of Carlos Castaneda's old Mexican Indian mentor, Don Juan, who says in *Journey to Ixtlan:* "What injures the spirit is having someone always on your back, beating you, telling you what to do and what not to do." In the Hebrew/Christian tradition the same wisdom is to be found in the story that the world's evil results from Lucifer, the angel of light ("conscience"), getting above himself and trying to be God.

The world has not yet found the pattern of family life or social order that achieves Lao-tse's goal, and so the topdog/underdog conflict remains a major life problem, and a major dream phenomenon, even in the most supposedly "permissive" circles. It is not confined to people like Sam who outwardly conform to the authority of their parents or teachers. Children who rebel and defy their parents or other moral authorities do not get away from their inner topdogs but pay a terrible price in guilt for their apparent freedom, as I know from my own painful experience.

My own main topdog derives from my mother, who instilled into me the ideal of cool, strong-willed detachment from all emotions and bodily feelings—sex, love of food or comfort, warmth, affection, anger, fear, or even enthusiastic joy. If I wanted to hug her, I was told as a child that everyone knew not to kiss her because she didn't like it; if I cried, I was told not to be a baby; if I lost my temper, I was told I was behaving like a monster; and when I was caught at age five in sexual explorations with the little boy next door, the resulting fuss shook the entire neighborhood for weeks. The whole situation was impossible, and though I tried hard to keep them back, my pent-up emotions would burst out from time to time —which just corroborated my mother's expectation of how awful they could be; she even suggested taking me to a "psychiatrist" to be cured. On leaving school, I rebelled, took off for Europe, and went on a jag of promiscuity, but the freedom was quite illusory, for my inner topdog always contrived somehow to spoil my fun.

What is more, the same futile seesawing between repression, followed by explosion, followed by guilt and more repression, on and on endlessly, continued long after my head had grasped that my mother's injunctions to her child were the result of her own emotionally deprived upbringing. It was only when I worked with my dreams in the context of Gestalt therapy that I learned how to end the self-torture game and know real freedom—which means freedom to express needs or hold them in check entirely according to the merits of the situation (the only legitimate controller), not by conformity to or rebellion against rigid rules dictated by either external or internal authority. When underdog breaks out slyly like Sam, or in some kind of overt rebellion like me, he very often goes too far, well beyond what the individual really wants, for the simple reason that he has been kept down for so long—but there is no way of knowing what his *real* needs are if topdog is constantly waiting in the wings to come out wagging his finger and saying, "There you are. I told you so. I warned you. This isn't what you want at all. Now if you'd only listened to me in the first place."

A major goal of all psychological therapies is to re-educate conscience so that we can accept and live in accordance with our natural human self, and Perls's aim is to get underdog to stand up to topdog honestly and challenge all his premises with conviction, which is something quite different from rebellion. Sam made a faint attempt at rebellion in his dialogue when he said, "All work and no play makes Jack a dull boy," and in waking life he was secretly planning that long weekend fishing. At points like this in a dialogue, Perls would often interject, "I hear you whining"—and he would repeat this even when underdog plucked up courage to rebel more aggressively, for blustering is only concealed whining. This actually happened in Sam's case, for at the point where he gave in to the nagging of "topdog car," the whole group roared for him to stand up for himself, and he made a valiant effort to do so. By this time, since it was obvious that the "car" of his dream was speaking with the voice of his father, we suggested that he put his father in the chair opposite and continue the dialogue with him, whereupon it went something like this:

Sam: Look, I'm sick and tired of your always getting on my back like this. You killed yourself with a heart attack, and I'm not going the same way. I'm grown up now and can make my own decisions, so stop making my life a misery.

At this point, as often happens, the group applauded and Sam looked pleased with himself, but we knew that victories against topdogs are not so easily won, so we asked Sam to move over and answer for his father.

Father: Now, Sam, you know I wouldn't nag you if it weren't for your own good. There's nothing wrong with fishing —wish I'd had time for it myself, but I was always too concerned about you and your mother for that—but with your promotion just around the corner, it's not the right time for it. Besides, you know perfectly well in your heart of hearts that if I didn't hang around reminding you of your work, you'd just forget all about it—and what would happen to your poor family then? We've got to have some controls in life, you know, or we'd just get out of hand.

Sam: Look, I've told you I'm big enough to take care of myself and make my own decisions . . . (voice trailing off and turning to the group) . . . but he's right. We do need some controls. It would be so easy for me sometimes just to sit there fishing and enjoying myself, and let my work slide . . . and that's irresponsible and uncaring. I couldn't live with myself if that happened.

This is the impasse or "sick point" where the energies of the personality become totally stymied—and those who pursue a course of rebellion against their topdogs suffer it just as much as those who give in. They cannot throw topdog into the garbage pail because they really feel that without his ministrations they would indeed become irresponsible, uncaring slobs, or whatever. To go through the impasse, then, is to risk losing one's precious self-image which, as Sam said, means death if you feel you can't live without it. This is why Perls saw a "death layer" waiting behind the impasse—the catastrophic expectation that if we express our needs and feelings, we shall be unloved, punished, persecuted, and we shall die, exactly as a child feels at the prospect of losing his parents' love. So rather than go through the impasse into the death layer, we implode our energies, constrict ourselves in a kind of catatonic paralysis, and live a half life at best. The first impasse occurs in childhood when it dawns on the small child that he will never be loved *for himself* but only for what his parents expect him to be, and he usually spends the rest of his life playing the roles that bring him love and approval. But there is a great difference between the plight of the child and that of the adult: whereas the child really has no resources to survive without his parents' care, the adult has, but because he still thinks of himself as a child, he does not realize this and remains at the impasse.

While Perls was all for people mobilizing their own resources and doing a certain amount of self-therapy, he saw the impasse as a point beyond which the individual cannot go without help, for the simple reason that he cannot see the obvious where another person can. This is why we suggest that people work together in groups or couples; but our own experience and that of countless other people shows that liberation is not out of the question for the dreamer working on his own. The step is an *intellectual exercise* in which one tries to *catch topdog out in a lie*—a lie so monstrous yet hidden that you have missed out on it throughout your life. As Sam was working in a group on his dream, I will demonstrate how the other members helped him to see this lie (we deviate from strict Gestalt therapy in allowing group members to share in the experience of the individual working on his dream); but please bear in mind that you can perfectly well do this on your own.

Sam had stopped at the point where he was in total agreement with topdog that "we must have some controls"—and it seemed that impasse had been reached. Sam could see no way out of the dilemma.

Group member: Yes, Sam, I agree too—we must have *some* controls . . .

Another group member: He's right. You can't fish *all* the time . . .

Sam: (looking puzzled) But I don't want to fish *all* the time, only some of the time when I feel like it. I enjoy my work, too, you know . . .

The group laughed and applauded, becoming silent as light dawned on Sam's face. "Wow," he said, leaning forward in his chair and facing "father." "You con man, you trickster . . . all my life you've implied that I'd be a total slob if it weren't for you. I don't want to fish *all* the time, you old bastard. I never did, and I never will—there's not the slightest danger of my becoming a slob. I see now what you've done to me—

you aren't advocating *some* controls at all—you're wanting complete domination and control of my entire life. Well, you can't have it. I won't let you—so get the hell out of my life. Get out, out, out!"

At this point Sam broke through the impasse with an explosion of energy that he later described as strange, terrible, and wonderful to him. We usually keep some cushions tucked away for these occasions (and tissues if the breakthrough involves crying) and we threw one for Sam to pummel as he released a lifetime's anger, rage, and frustration on his father. My own anger against my mother was released in a similar way when I realized that all my life I had bought her notion that without her total control I would become merely a messy slush of emotion. (My own concern about my very passive father came out in another group I attended when the leader asked me why I was trying to live his life for him. I replied, "Well, *someone* has to," at which we all burst out laughing at my unquestioned assumption that he was incapable of doing this for himself!) The other realization which follows is that the people we are trying so hard to emulate are themselves often sorry figures with little life of their own—but the amazing thing is that even those of us who consciously decide we want to be quite different from our parents still go on carrying in our minds the injunctions and catastrophic expectations we learned from them, inevitably ending up as sad replicas of them against our will. When we break through the impasse all the imploded energy we have been storing in our bodies as anxiety explodes into excitement, which is the very energy and essence of life. The result is often a mini-satori, as Perls described it, in which a robotized corpse comes to life.

However, this is rarely the end of the story, for topdog can be a tricky customer when his cherished authority is challenged: we may think we have thrown him out of our lives, only to find him creeping in through the back door. A typical case in point was a girl in one of our workshops who dreamed that one of her children felt neglected when the real child had no such feelings (or none that she did not already know about). By acting the part of the child, who put on a brave face in coping with life's difficulties, including his father's death at an early age, she got in touch with her own need for love and comfort lying behind her stoical exterior in coping with the death of her own father in childhood, and later of her husband. This was a revelation to her at the time, and the group members who knew her well resonated to this hidden need in her—but the following week she returned and said that she couldn't possibly allow herself to fall into self-pity. We laughed and asked who said anything about self-pity—and she realized that topdog had crept back, trying to seduce her with his old subtle lies and catastrophic expectations about what would happen if he relinquished control.

In reality, the very fact that topdog's injunctions appeal to us at all means that his virtues are *already there,* firmly implanted in our being, and will emerge in our growth without our having to struggle and strain away at them as if they had to be carved out of some wholly recalcitrant material. In the case of parental topdogs, many of their real virtues are probably quite literally in our genes—but of course we have other sides to our natures as well, and the only realistic way for us to grow is as a harmony of all our characteristics. This is usually not enough for parents and authority figures, however, who tend to be more interested in our obedience than in our growth, and this is even more the case with our inner topdogs. Like Lucifer, they are not content to be parts of the greater whole, serving the total economy of the personality by bringing their own special gifts to it when needed. They want to rule and will make life hell for us in order to achieve this, like Satan, the Father of Lies, in Milton's *Paradise Lost* whose motto was "Better to rule in hell than serve in heaven."

The real secret of getting free from our little private hells is not to try to throw our topdogs into the garbage pail altogether—this cannot be done as they are a part of us—but to stop their tyrannical rule over us, knowing that we *already* possess within us the virtues they extol. That is why, in our workshops, we try to bring about a creative resolution by asking both topdog and underdog what they appreciate about each other and how they are going to function as a whole (a Gestalt) in the future. The conclusion of Sam's dialogue with his inner father went something like this:

Sam: Well now, you old fraud—get this straight. There are a lot of things I appreciate about you—your concern, your need for some kind of higher life, your perseverance—but you've been going about it in the wrong way. You've sabotaged your own efforts. I shall be much more like the person you want me to be if you leave me alone and stop nagging. You might just serve some useful purpose if you become an adviser rather than a ruler—in fact, you can have the job and nudge me from time to time if I ever overstep the mark.

Father: O.K., Sam, you win. And if I'm honest, I have to say I admire you for it. I never did like doormats, and I always hoped you'd stand up for yourself one day. And I admire your ability to relax and take things easy—I never could unwind and it killed me. And you've made a relationship with Ellie that I never managed to make with your mother. I'll hang around—I mean, I wouldn't have stayed around for so long if I didn't like you—and I'll watch you grow, doing the odd bit of pruning and fertilizing if you need it. You know, it's kind of nice not having to bother about you any more —guess the first thing I'll do is take a rest. . . .

The outcome is not a compromise, for in a compromise each side loses out. It is a creative resolution in which both sides win because both get everything *real* that they want. In practical terms, two loads of energy are released for creative living —that formerly dissipated in the self-torture game, and that actually kept down in underdog himself. All this is now available to Sam, who will find that he gets everything done in half the time it took previously, and so will have ample time for fishing. (We later heard from Sam's wife that he never again carried his briefcase full of work on vacation, much to her relief and pleasure.) The only thing lost is topdog's rule over the kingdom of Sam's personality, but his real virtues show more than they ever did. Stripped of his usurped power, Lucifer reverts to his proper role as guardian angel or angel of light. In this new role you can even call him conscience if you like, so long as you remember always that conscience or guardian angel is like fire—a good servant but a terrible master. There is a principle of "rightness" in us which is altogether higher than conscience, for conscience even at its best is always socially conditioned. The true divine principle is one of wholeness at the center of our being which causes us to grow naturally and function in a self-regulating system where "self-improvement" is redundant. This means that there is far more divine energy in underdog than in topdog. As Jung said, the "shadow" side of the personality which we normally try to keep down in the interests of some "higher" good is ninety percent pure gold.

The ingrained habits of a lifetime are unlikely to be totally broken in a single experience, however dramatic, so the achievement of a creative resolution like Sam's is more likely to herald the beginning of a new life-style that has to be consolidated over months or even years than a once-and-for-all breakthrough. Topdogs keep trying to reassert themselves, and if we watch our dreams, we can get to know when this is happening. But a really vivid dream dialogue like Sam's stays in the memory as a constant reminder of topdog's lies, and if we go on firmly putting him in his proper place, underdog eventually learns not to be sly but to stand up for himself responsibly, which means of course that he ceases to be underdog. This is the emergence of the true god principle in the personality.

Another point to remember is that most of us have more than one topdog inside us, corresponding to the fact that we have had several authority figures to cope with in our early years. Sam, for example, in addition to his "Puritan-ethic topdog," in a later group confronted what might be called a "financial topdog" who nagged him in waking life and appeared in his dreams whenever it felt he had spent money on something "frivolous." It took Sam a little time to learn that it was all right to allow himself a little frivolity on occasions without having to apologize for it or plead that it really served a useful purpose! I myself had to cope with a "Christian-suffering topdog" and an "academic-excellence topdog," as well as the sexual and emotional topdogs derived from my mother, plus several minor ones. So it is important to watch your dreams for the emergence of these clowns who are spoiling your life, catch them out in their lies, and terminate their self-torture games once and for all.

I would love to present a full dog show with all the different breeds of topdog that have emerged in our dream groups on display, but space restricts my choice at this point to only one or two, though more will appear in later chapters. Whereas Sam was identified with underdog in his dream, Mike in the following example is an onlooker at the self-torture game actually taking place on the stage of his dream.

A MARITAL TOPDOG—WITH UNDERDOG AT THE END OF HIS TETHER

Mike, a musician with three teenage children, had left the family home to live with one of his students a couple of years before the dream. He had subsequently returned to his wife, left home again, and was once more with his family when we visited him. This shuttle between wife and mistress had resulted in frustration and anger for all concerned, including his children, because no one knew where they were with him. Mike himself described his behavior in Perls's own terms as "neither shitting nor getting off the pot," a perfect example of topdog and underdog at impasse. A week before our arrival, Mike dreamed:

> I am in Miami on a bright sunny day, at the side of the Expressway, the elevated part. Together with some spectators I am looking over a barricade across the east side of the road, where traffic has been blocked, to a center part where a large tractor-trailer is upside down, a lot of splinters and rumpling of the body. It has been a horrible accident, and I am morbidly interested. With the other spectators, talk begins about how dangerous the expressway is, and I say, "You can drive it, I have, but carefully."
>
> I look and see a police officer with a grim face kneeling beside what looks like a tarpaulin. I look more carefully; he seems to be coiling and uncoiling black hose, like fire hose. I jump over the parapet, cross the highway and get close. It is horrible. The driver of the truck is unconscious, wrapped in what looks like a wet suit, or better, a flying suit, but it is wet with brine because the driver has been horribly burned. His face and head are in plastic, as if oxygen is being provided. But below the chest, his body is split open . . . his abdomen is spilling black intestines out. The policeman keeps straighten-

> ing them out, coiling them, putting them back into the body cavity. But they keep ballooning out as if they have a life of their own. Besides horror, I admire the policeman for his discipline and care.

Mike said he felt sure the dream was saying something about the disaster afflicting his marriage. He associated the conversation about the dangers of the expressway with the fact that when people ask him how one goes about being married and raising children, he often replies, "With difficulty," an analogy with the old riddle about how porcupines mate—"very carefully." He is also fond of describing Miami as "moral fiberglass," meaning a place of phoniness, so the dream seemed a clear picture of a crash on the expressway of his marriage caused by lack of moral fiber in ratting on his family. Mike's interpretation in his own words is as follows:

> I am an onlooker (in retrospect) because I have been burned, I have spilled my guts, and am in the process of dissociating myself (with difficulty) from the disaster, getting/pulling myself together. To me, the exposed guts which have spilled out mean that I have made myself absolutely vulnerable to Kate (mistress); now (in the dream) I am trying to put myself together to get back on the road to life.

We talked about the dream with Mike at odd intervals during the day, and he agreed that here was an actual picture of a topdog/underdog conflict in the dream itself. Policemen and similar authority figures usually symbolize topdogs, while the gut was behaving in a typically underdog fashion, slipping out of control despite every effort of topdog to contain it. While Mike was obviously on the side of the policeman in the dream, approving his careful and disciplined action in trying to push the gut back inside, further work on the dream, which went something like this, enabled him to get in touch with the feeling of the gut itself—that portion of himself which Perls would describe as "alienated" or unacceptable:

Policeman: Come on now, there's a good gut, go back where you belong. When you spill out like this, you're disgusting—look, you're black and full of shit. It's not nice for people to see you—it upsets them. Go back inside, please, where you're contained, comfortable, and can function harmoniously with the rest of the body.

Gut: But I'm *not* comfortable in there, and I wasn't functioning harmoniously. I'm too big to stay cramped in there—that's why I spilled out. I need more space and freedom, though I must say it's a bit scary not being contained at all . . . and I resent your statement that I'm full of shit—the blackness is the deadness I feel from being shut up for so long in too small a space.

Pol: Now, come on, you know very well that a gut *ought* to be inside—if you stay out, this poor fellow here will die.

Gut: Oh dear, yes, I suppose you're right, guts aren't meant to spill out, and I really don't want to hurt anyone.

This was the dynamic that had driven Mike back home each time he had tried to "break out"—and, of course, it would indeed have meant death to Mike's image of himself as the kind, loving, caring family man had he decided to stay out. The dream made it clear, however, that underdog had reached the end of its tether, and the time had come to stand up to topdog something like this:

Gut: Look, if I *could* go back I would, but I *can't.* There just isn't enough space for me inside. I'm too big for it.

Pol: Nonsense, you're not really trying. Say "I *won't*" instead of "I *can't.*" It's just lack of moral fiber. If you constricted yourself enough you could do it.

Gut: Right, you asked for it. I *won't* go back inside, I *won't* constrict myself so that I cease to function, I *won't* tie myself in knots trying to do what's good for other people all the time . . . quite apart from the fact that

I was doing nothing but harm in that dead state. If I stay out, it's death to this chap, and if I go back in, it's death to me—which means death for everyone.

Pol: I shall call the surgeon to sew you up so you can't come out.

Gut: Then I shall just cause a hernia. I'm not going back, I tell you . . . we have to find some other solution.

Pol: But there isn't one.

Gut: Yes, there is. How about enlarging this fellow's body to make room for me? That way he stands a chance of living, but if I keep bursting out, he's sure to die.

Pol: All right, you win. And you know something, I rather admire your determination to stick out for your own needs. It takes *real* guts to do that.

Gut: And I appreciate your concern for this fellow here and for me, too. You're right, I do need to be contained in *something* or I just make a mess, but it has to be bigger. But you couldn't know this. It might have worked with someone else. What's right for one person isn't necessarily right for everyone.

Mike had no difficulty getting the message—that he just had to find a larger, freer life-style than his present marriage. But more important, Mike realized he also had to make room in his own inner space for the *feeling* part of his nature that he had been putting down all his life in the interests of reason and intellect. Because he had constantly tried to control his feelings in all aspects of life, they tended to balloon out when he least expected it, and he had probably made himself more vulnerable to Kate than he really wanted. Only by letting go the self-image of the kind, reasonable, caring family man who had life sewn up, could he resurrect to something altogether bigger and better, thus avoiding further psychological hernia and possibly a physical one. He is now living in an "open marriage" situation with his wife, and they are both making new lives for themselves without the cramping mutual dependency of traditional marriage. He feels far more at peace with his family than he ever did before, so the "policeman's" concern for their welfare is being met far more effectively than it was when he tried to force Mike to conform to the old pattern. Mike wrote recently that he has no idea whether he will stay married or not—he will allow the situation, not a rule, to decide. He added in a recent letter:

> I wonder with some sadness about Kate, and how she is doing . . . I have not contacted her, nor she me. I am looking forward to new adventures, new loves, and I feel far freer than I have in years. And I see my wife getting more so as well. Who knows—I may, like Zorba, loosen my belt, and run headlong into another woman, only this time I shall feel less encumbered. Anyway, it's good to be alive! I love me, among other people, and think I'm magnificent. And modest!

Mike's problem is such a common one that we must welcome the way many people today are experimenting with alternatives to the traditional pattern of family life based on monogamous marriage. So long as we retain the social assumption that this pattern is the right and proper one for everyone (an extremely improbable notion in view of the wide diversity of human character), then no amount of liberalization of divorce laws or sexual morality will prevent people from feeling topdogged by the knowledge that their dependents are counting on them to stick to the pattern however much it goes against the grain of their essential being. It does indeed take real guts to break out when your inner voice is saying, like Mike's, "It's not just yourself and your own needs you have to think of: the lives of other people are at stake, so try a bit more willpower." The concealed lie here is the failure to realize that partners and children will suffer even more from living with someone who kills himself inwardly by putting down his needs (or tries to do so only to keep breaking out, repenting, and breaking out again). All patterns of living involve *some* discipline, of course, but when the pain is as great as Mike's, no discipline in the world can make good come of it—as everyone realized in the end.

Whatever the cost of Mike's decision to the others around him, it will certainly be a great deal less than the cost of any alternative course of action based on trying to put down his needs. As a matter of fact, none of his dire expectations have been realized, and it sounds as though everyone is better off than they were before. If we can learn by experiment to create a society which offers many different patterns of living together—group marriages, communes and monogamous marriages with or without sexual freedom, and so on—so that children grow up with no fixed patterns of expectation about the "proper" way for needs to be met, we might hope to reduce the number of cases where problems like Mike's arise, perhaps even abolish them altogether. Religious diehards and others who think this would mean social chaos are ignoring the fact that chaos exists already. The traditional pattern is collapsing, not from lack of willpower, but because we are no longer willing to be hypocritical about the pain that has always festered beneath it. Those who feel no need for alternative options should remember the words of William Blake before trying to push their trip on others:

> Those who restrain desire, do so because theirs is weak enough to be restrained; and the restrainer or reason usurps its place & governs the unwilling.

For a creature as complex as man, the world has to offer as many options as possible, for there can be no simple, ideal life pattern for everyone, nor even for the same person at different stages of his life.

THE BULLDOG BREED

Perls once remarked that our difficulty in getting free of our topdogs may be greater than it need be because we have a rather sinister motive for hanging on to them. In *Gestalt Therapy Verbatim,* he states:

> We usually take for granted that the topdog is right, and in many cases the topdog makes impossible perfectionist demands. So if you are cursed with perfectionism, then you are absolutely sunk. This ideal is a yardstick which always gives you the opportunity to browbeat yourself, to berate yourself and others. Since this ideal is an impossibility, you can never live up to it. The perfectionist is not in love with his wife. He is in love with his ideal, and he demands from his wife that she should fit in this Procrustes bed of his expectations, and he blames her if she does not fit. . . . the essence of the ideal is that it is impossible, unobtainable, just a good opportunity to control, to swing the whip. The other day I had a talk with a friend of mine and I told her, "Please get this into your nut: mistakes are no sins," and she wasn't half as relieved as I thought she would be. Then I realized, if mistakes are not a sin any more, how can she castigate others who make mistakes?

This seems to explain the very puzzling behavior of a certain topdog of the bulldog breed, known as the British topdog. This particular topdog scorns and despises the materialism, brashness, vulgarity, lack of taste, and commercialism of his American cousins and wastes no time in condemning the "superficial" American way of life—but he is found from time to time on this side of the Atlantic, giving lectures or fixing profitable business deals whenever he finds himself short of cash. The underlying dynamic came out when an academic friend of ours came to stay with us in Florida at the end of a six-week lecture tour, most of which, he said, had been a terrible chore on account of the noise and superficiality, but he had been forced to do it for the money to send his daughter through college in England. It would have taken him almost a year to earn the same sum of money back home, and while he was at pains to impress upon us that full taxes would be deducted, he clearly could not get out of his *heart* that the whole venture was a dishonest and unforgivable sin on his part, as his dream the first night showed.

In it he was watching with distaste and disapproval as an American in a big, flashy car backed down a driveway and cut across a corner of the yard, making rather a mess. Telling us the dream over breakfast, he commented that this was how he saw the Americans—all in such a hurry to make money that they "cut corners," heedless of what they spoil in the process. Since he was rushing to catch a plane, we had no time to do more than remind him that dreams never come to tell us what we know already, but we could not help laughing after his departure at the fact that he had, the previous evening, justified his lucrative lecture tour of the U.S. as a case of "spoiling the Egyptians," as the Israelites were told to do in the Bible.

Had he been staying on with us, we would have suggested that since the dream told him nothing new about his feelings for the Americans, he might try exploring the possibility that it depicted *himself* cutting off a very large corner in earning enough for his daughter's education in a very short period of time. The driver of the car would have symbolized an "ugly American" underdog inside himself, despised by his tasteful, uncommercial, sensitive British topdog, yet finding an excuse nonetheless for doing what it wanted. A dialogue between our friend, X, and the driver might have gone something like this:

X: There you go again, dashing off to make money. Typical! You're in such a hurry that you spoil everything. You should be ashamed of yourself.

D: But I have to have this money. I've got family responsibilities.

X: That's no excuse for spoiling things. You've made a terrible mess. You're so insensitive.

D: Oh dear, yes, I didn't realize . . . I didn't notice. I'll take more care next time.

X: I sincerely hope there won't be a next time. You don't need this extra money. You have two homes already, a very comfortable job, and you can well afford to send your daughter through college without dashing off on these wild-goose chases.

D: Well, I suppose you're right—but I like it this way. I enjoy going off sometimes—it's not just the money, it's the whole scene. I get pretty bored just staying at home all the time, or doing my comfortable job. I need a bit of excitement from time to time.

X: Well, all I can say is that it's very immature of you, and you should do your best to correct this childish impulse.

D: Yes, I know. I'll try not to be weak again. (*Thinks:* I'll find a way.)

As I write this, I can't help laughing because, although the dialogue comes out of my own fantasy of how my friend would do it, I would bet a large sum of money that this is a close replica of the conversation that took place between our friend and his wife before his trip. In a letter to us before his visit she wrote that she preferred to stay at home in the peace and quiet of her English country garden and watch the bulbs grow, implying that her husband was an immature, insensitive adolescent to prefer the noise and vulgarity of the great American scene. It is easy to imagine him pleading in his turn that they need the money. I now suspect what had not been clear to me before, namely that X was dying to visit the States, a change from his in-turned, quiet, and undramatic British existence (he had just come to us from California, where his famous friends had given him the full red-carpet treatment), but could not bring himself to admit it to his wife—and more importantly, to himself.

International observers have written a great deal in recent years about the "neurosis" that seems to have affected the British economy since World War II. I feel sure that one aspect of it is the refusal of the British to acknowledge their own desire for money, success, and excitement, out of the expectation that to do so would immediately result in the whole of Britain being overtaken by the very worst aspects of American commercialism. The bulldog breed would be a whole lot healthier, as well as less hypocritical, if they came to terms with the reality of their own feelings—but then, of course, they would have no basis left for criticizing the Americans, and the last remaining bastion of the British sense of superiority would fall!

UNDERDOGS TURNED TOPDOG

I have met attitudes similar to those of the British topdog in some American academics, but here quite a different dynamic is usually at work. While the British attitude of lofty superiority to "brash self-assertion" and "blatant materialism" is a social pattern going back several generations and firmly built into the national educational system, Americans who adopt this attitude are normally, in my experience, rebelling *against* the dominant values of the society in which they were brought up—and this introduces a very important general point.

Because underdog (so long as he remains underdog) accepts the sadistic moralizing premises of topdog and tries to answer back in kind (e.g. meeting cliché with cliché), people who are underdogs in society can generate very nasty new topdogs if they rebel successfully and try to create a life-style based on their rebellion without really working through the problem inside themselves. On the political plane this happened, with horrible results, in both the Nazi and the Communist revolutions, where those who had been society's underdogs rebelled against an order that gave them no legitimate outlet for their energies and then proceeded to impose their own values and prejudices even more harshly than the old order had done. Probably the most outstanding example in history, however, is Christianity, which has instilled killjoy topdogs into generation after generation, on the basis of statements made by Jesus in his own time to try to lift up the hearts of the poor, the suffering, the oppressed, and the sexually exploited.

Christian topdogs are of many kinds and are particularly hard to see through precisely because Jesus's ethical values were in general so obviously admirable. William Blake was so impressed by this phenomenon that he said most Christians had fallen into the trap of worshiping Satan under the name of Jesus. Once the essential principle of the topdog/underdog self-torture game has been understood, however, it is possible to catch out topdogs who sermonize in Christian terms about the need for humility, unselfishness, poverty, chastity, suffering, or whatever, in exactly the same way as any other topdog. Everywhere and always, the topdog lie is the same, namely that the good life is to be achieved by allowing ourselves to be *controlled* by the inner voice that nags us to do things we don't really want to do at all. As Blake said, if this had been what Jesus was after, he would have been a Pharisee, not someone the Pharisees hated. If the values of Jesus appeal to us at all, then they will emerge naturally in their own good time in their own good way according to our own basic growth pattern. Any attempt to achieve them by force or imposition will only bring about the opposite result. As Perls says:

> We are all concerned with the idea of change, and most people go about it by making programs. They want to change. "I should be like this" and so on and so on. What happens is that the idea of deliberate change *never, never, never* functions. As soon as you say, "I want to change"—make a program—a counter-force is created that prevents you from change. Changes are taking place by themselves. If you go deeper into what you *are*, if you accept what is there, then a change automatically occurs by itself. This is the paradox of change.

In our work with students we have become very much aware of a new underdog-turned-topdog rearing its ugly head among those who think of themselves as members or supporters of the counterculture. This too is a topdog whose evil identity is often hard to detect because he says so many things that are obviously right. It is true that Western society is often corrupt, greedy, and inhumanly exploitive; it is true that our scientific world view is often life-denying in its narrow materialism, and our technology destructive of the environment; most important of all, it is true that we badly need to develop alternative ways of life for those who find that the "straight" world denies them the opportunity for growth and essential self-expression. But as soon as counterculturally minded people allow themselves to moralize about these things, preaching that everyone *should* be peaceful and gentle, *should* be warm and feeling rather than intellectual, *should* share with others, *should* never go on an ego trip, *should* never make money, and so on, they fall into the very worst evil of straight society—which must inevitably produce a new generation of hang-ups.

I detected this counterculture topdog at work in myself one day when I had been looking at rather expensive waterfront houses. That night I dreamed I was flying, but my counterculture topdog forced me down by giving me an injection (an anti-bloated-capitalist shot, I suppose). I then found myself on a hillside where young people were sitting around playing flutes and sewing. One girl said she would make me a dress for $200. Just as I was thinking that this was a very expensive dress, a voice in my ear whispered, "but *she* needs it," implying that the counterculture needs my money more than I do. On waking and understanding the dream, the most I could answer in a most underdoggy way was, "Yes, you're quite right." But somewhere deep inside I was furious at the encroachment on my own innate integrity by a topdog who believed I had to be bulldozed into using my money in the right way. I learned my lesson and yelled at him, "You get the hell out of my life. Don't you preach at me. I know all about your new values and *I'll* decide when to use them. I'm perfectly capable of using my money sensibly. In any case, the more you nag me, the more likely you are to lose your case, so get out of my life."

Inner topdogs never come out of the blue: they have to originate somewhere with a real voice. I had absorbed this one from the high moral tone of many counterculture spokesmen. Many parents will recognize the same hectoring note in the voices of their children, and in our dream groups we have found many young people completely twisted up by the counterculture topdog inside themselves forbidding them any outlet for their natural energies and ambitions. My advice to the counterculture is to remember how they felt about straight society's topdogging and try to learn the psychological lesson instead of merely swinging the pendulum to the other side. As California Assemblyman John Vasconellos wrote in the *Journal of Humanistic Psychology,* politics must take account of inner realities:

> Increasingly I come to realize that the discovery of a new politics for our culture depends upon my living and experienc-

ing and discovering a "new politics" within myself—getting so much in touch with all the parts of my own being, that out of the resultant oneness within me I will increasingly live disclosingly so as to expose the institutions and customs of our culture which stand in the way of oneness, within ourselves, between ourselves, between us and the earth.

FRITZ IN THE MANGER

As this and following chapters demonstrate, I believe Perls has made a major contribution to psychology and psychotherapy by giving us the "topdog/underdog" model, which anyone can use to break out of the self-torture game. While this may not be the end-all of therapy or psychological growth, it is certainly the first essential step in clearing out our inner space and getting ourselves together, for so long as we are at war within the personality, we shall be at war with the world. Even those who are undertaking religious or spiritual growth disciplines should bear in mind that you cannot take half of yourself to God—it has to be all or nothing, and we cannot encounter the Hound of Heaven until we are free of the futile dogfights of the Hounds of Hell.

Because I value Gestalt therapy so highly, it hurts me to find that it often becomes a new dogma in its turn, putting down all other approaches to therapy and to dreams. For example, Perls insisted that *all* dream images are necessarily parts of yourself, representing alienated elements of the personality which you have pushed away because you do not like them. As I have tried to show throughout this book, at the risk of being repetitious, a dream can be understood on *many different levels,* and often the beauty and precision of a dream, with a clear message, emerges without going as far as holding dialogues between the characters. To insist on looking for topdog/underdog conflicts in all dreams, even in all "inward-looking" dreams, is to push the river (Perls's own phrase) just as much as psychoanalysts who are dissatisfied until they have reduced a dream to sexual terms.

To try to cram all dreams into any limited mold or theory is to join the ranks of the dream-killers. In our own work we try to let the dream unfold naturally, using whatever means are suitable for the situation, and certainly allowing interpretation when appropriate. Again and again, however, we encounter Gestalt therapy "top-dogmatists" yapping that any attempt to look for meaning in a dream is "mind-fucking," avoidance of the true emotional issues.

Of course, a Gestalt therapist can argue that his concern is to use the dream for therapy in the way he knows best, rather than to discuss the dream itself—which is fair enough provided he admits it and does *not* try to put down other insights from dreams. Perls's model of the integrated personality is distinguished by its emphasis on expansion of consciousness to embrace even "unacceptable" parts of the self; it is a pity his followers do not take the same inclusive approach to different ways of using dreams. So if you ever go to a Gestalt group and find the leader top-dogmatizing about dreams, tell him you know better, and if necessary repeat to him Perls's own Gestalt prayer:

> I do my thing, and you do your thing.
> I am not in this world to live up to your expectations
> And you are not in this world to live up to mine.
> You are you and I am I,
> And if by chance we find each other, it's beautiful.
> If not, it can't be helped.

«10»

THE SECRET SABOTEURS

> As I was going up the stair
> I met a man who wasn't there,
> He wasn't there again today—
> I wish, I wish he'd go away.
>
> —Hughes Mearns

We first detected a hidden villain behind the events of a dream in a student workshop when Amy described a dream about getting a parcel of music from home but being continually thwarted in her efforts to find a piano on which to play it. When at last she found one, her music was missing, and when she retrieved it, the piano was occupied; and sitting down at the next piano, she discovered it was out of order. This vicious circle continued until Amy woke up crying with rage and frustration at being continually sabotaged in her attempts to play.

Amy told us that she had, in fact, received her music from home the previous day and was indeed looking forward to playing it. When we asked why she had written this frustrating dreamscript for herself, she was at a loss to answer, for she could not resonate at all to the notion that she really didn't want to play. However, she accepted that at some level she must be in two minds about it, and so we asked her to talk to the Amy who did not want to play the piano—the secret saboteur who never actually appeared in the dream itself, but who carefully arranged that she should forget her music and that the piano should be occupied or out of order. A simplified version of the dialogue went something like this:

Amy: Why are you trying to stop me playing? I've really missed it since coming to college, and was so much looking forward to playing again.

Saboteur: You know very well you shouldn't be playing this music. It's a total escape into yourself when you know you should be out making friends and learning to socialize. I'm doing it for your own good. If you're going to live in the world, you've got to mix with people, talk to them, go out to them. If you don't learn now, you'll regret it later on. Believe me, I know what's best for you.

Amy: It's not an escape, Dad.

At this point both Amy and the group burst into laughter as the identity of the hidden villain was revealed. From the dialogue it sounded as if this particular conversation had taken place many times before, and Amy corroborated this. In fact, she had deliberately left her music at home on coming to college so that she would be more or less forced to turn outward in an effort to live up to her father's expectations of what a healthy young girl should be. She had missed it so much, however, that she had asked her parents to send it on. It was as though her father had pinned a note to it, saying, "Here's your music. You see, I'm not the wicked parent trying to deprive you of your little pleasures. But every time you play it, just remember what I said," which made Amy so guilt-ridden that she would eventually put the music on one side and forget about it.

We asked Amy to put her father in the opposite chair and stand up to him. The dialogue went something like this:

Amy: My music isn't an escape, Dad. I've tried to tell you this before but you never listen to me. And here at college, I have made friends and I am much more outgoing—but I need my music too. You've never understood that I really love it. [At this point, Amy's face lit up with such rapture that we had no doubt that she was telling the truth.] What's more, I don't want to be totally extroverted. I think that's a shallow way of living. I'm not one of your neurotic patients, you know, who can't cope with the world. I'm Amy, and I'm not a child any more. I'll make my own decisions in future. I don't need you on my back all the time telling me what to do and how I should be. You are you, and I am I. [Amy burst into laughter, saying that her father had the Gestalt Prayer on the wall above his desk!] Look, Dad, read your poster and leave me alone . . . I know you love me and think you're helping me, but you're not. So don't ever try to stop me playing again.

Dreams in which circumstances mysteriously conspire to frustrate us are extremely common, and our work with Amy demonstrated an entirely new approach to understanding them. Instead of making unhelpful and self-evident statements like "you don't *really* want to play"—an observation often made by wiseacre friends with a slight knowledge of dream psychology and also by professional therapists—we set out to discover the identity of the hidden villain intent on sabotaging our plans, thereby bringing the conflict right out into the open. In this chapter, I shall show how we have used this approach to probe the hidden dynamics of several very common frustration themes, including those of missing a vehicle, failing to get through on the telephone, finding that a door, stairway, or road has vanished, forgetting lines or inability to perform, and being unprepared for some important event.

We soon discovered, when we began to use the secret-saboteur approach, that there can be underdog villains as well as topdogs like Amy's. In fact, in some ways underdog is even more sinister when he adopts this concealed approach than topdog is. When topdog frustrates us, it means that somewhere in life guilt feelings are spoiling our fun by preventing us from doing what we want to, or making us feel so bad about it that we get nothing out of it. When underdog frustrates us, on the other hand, it means that we have been trying so hard to live up to topdog's expectations that we have been putting down our real needs. If we continue to deny underdog his rights, he will take really drastic steps to get his way covertly, like manipulating others into doing his will, and even going to the length of falling sick, for underdog is more closely in touch with the body's energies than topdog.

A good example of a hidden underdog was given to me by twelve-year-old Kathryn, who dreamed she was at school looking for her locker, number 814, but could find only numbers 813 and 815. Hers was missing. Kathryn wrote in a letter to me, "I interpreted the dream as meaning that since my locker and books weren't there, I didn't want to be there at all. I think school is O.K. but it could be a lot better." Clearly Kathryn's underdog is not so balanced in his feelings—he finds school so bad that he will go to the length of removing her locker and books so that she *cannot* go. I don't know the outcome, but I feel that unless Kathryn starts to find school more satisfying, her underdog could find a way to prevent her going there in waking life, perhaps by making her fall sick or have an accident.

MISSING THE BUS, TRAIN, BOAT, OR PLANE

Dreams of missing a vehicle can, of course, be warnings on the simplest level to allow yourself enough time to prepare for some event or opportunity in the near future, especially if you are normally an unpunctual person and the event is important to you. If there is no such event in your life, then the dream may be a simple pun expressing your heart's thought that you are somehow "missing the bus" or "missing the boat" in life, which could be anything from a promotion or new relationship to the feeling that you are missing out on life generally. The degree of annoyance or worry or grief you experience in the dream reflects your heart's degree of regret about the missed opportunity, and this can be quite revealing. For instance, if you feel less annoyance at seeing the vehicle disappearing in your dream than you would expect, you may have to wake up to the fact that you are really not as bothered about the lack of promotion or whatever as your waking mind thinks you are.

However, if the dream is one of those where events conspire to make you miss a vehicle, and you can see that it points to your being of two minds about something in your life, then you must unmask the secret saboteur who drove off just as you arrived or arranged the traffic jam on your way to the airport. This is especially true if such dreams recur: for example, John used to have frequent dreams of missing trains home after he had been out lecturing and so on, and being left lonely, cold, and miserable on the station platform. Some of these dreams occurred after he had actually been out the previous day but often he was unable to relate them to a day's event at all. So the last time one of these dreams occurred, he put the hidden villain who wrote the frustrating dream script into a chair, and the dialogue went something like this:

John: Why have you made me miss the train? I've been out working hard, I'm tired, and I want to go home. Most uncivil of you to play a dirty trick like that.

Saboteur: Not at all, my dear fellow. I'm doing it for your own good. I don't know how many times I've told you to give up this silly ego trip of lecturing and broadcasting out of office hours when you could be enjoying the true deep pleasures of hearth and home. You won't listen to me, but deep down you know I'm right. I just don't understand you. You must be suffering from brain damage or something—so I have to take matters into my own hands to teach you a lesson, to bring you to your senses. If I punish you enough for going off on these trips, by making you miss trains and leaving you cold and lonely on platforms, you may just decide to do as I say next time and stay at home. Now, for the next few hours you can sit here, think over what I've said, and repent of your past immature behavior.

John: That's very unfair. I *want* to go home—I missed the cocktail party after my lecture especially to catch this train, and I can't go back now, it's too late.

Sab: Serves you right. Now take your punishment like a man.

John: Look, I don't do this lecturing for pleasure. We need the money.

Sab: Rubbish. I saw through that old story long ago. Of course, everyone can use a bit of extra cash, but that's not why you accept these invitations and you know it. You just get this silly immature itch to be out in the big wide world meeting important people—and the time has come when you just have to grow up.

John: Oh dear, you're quite right—I won't do it again. (*Thinks:* I've got to shut him up for now, but I'll find a way to get round this one.)

The secret saboteur stood revealed as a "home-and-hearth topdog," probably derived from John's long-dead mother, but now speaking in tones reminiscent of his first wife, his church friends, and a number of psychoanalysts, all of whom preached the supreme virtues of warm, deep, personal relationships and the vanity of "superficial" academic or intellectual pursuits. In fact, when John's first marriage temporarily collapsed many years earlier, all these people had blamed it on John's "immaturity" and exhorted him to pay more attention to family needs. For a while he had complied, but since he really loves lecturing and broadcasting, and enjoys intellectual conversation, John's underdog had used the excuse of needing money in order to do these things, despite his conviction of topdog's essential rightness. But he paid the price of guilt when topdog's voice pursued him, constantly yapping that he was exchanging his true human birthright for a mess of pottage.

The home-and-hearth topdog is a very common one in our society and was brought to the forefront of public attention when Betty Friedan demonstrated its disastrous effect on women in her book *The Feminine Mystique.* Scarcely anyone realized that men often suffer from it too, however, until Esther Vilar produced her book, *The Manipulated Man.* Here is a classic case of one of society's underdogs (women) turning the tables, and becoming a new tyrant of terrifying proportions instead of abolishing tyranny. While the traditional feminists sought to secure equal opportunities for women in the commercial, professional, and intellectual worlds, a backdoor movement was taking place in which those women who accepted the domestic role were subtly instilling into the minds of their sons a feeling that all traditional "masculine" pursuits were crude, insensitive, and morally inferior to the "warm, fully human" values of home and deep personal relationships.

This trend has been decidedly reinforced by the spread of the psychotherapeutic movement extolling the virtues of feeling and sensitivity. A great deal has been written by feminists about the damaging male chauvinism of psychoanalysis, with its theory that women are condemned by their anatomy to find fulfillment in subservience to men, and I suffered much from this in my own analysis—but there is another side to the story. More modern schools of psychotherapy, and even many psychoanalysts, have laid such emphasis on the "immaturity" of sex without total, loving relatedness, and on the "adolescent escapism" of intellectual and organizational interests, that a whole generation of men has emerged who feel like John. When John confronted his topdog in dialogue, by this time personified by a psychoanalyst called Paul, the topdog's hypnotic spell was revealed thus:

John: Look, Paul, you're being a bit hard. I enjoy my home, and I don't go away too often.

Paul: My dear fellow, that's not the point. It's the fact that you're always *hankering* after these absurd ego trips. And you never learn, even when you get bored at academic meetings and cocktail parties. If only you'd listen to me.

John: Well, yes, they are boring sometimes, and then I wish I'd stayed at home. But, you know, I often get bored at home too, and then I wish I were out doing something more exciting. I love the family, but I love other things too.

Paul: You're just trying to escape the responsibility of intimacy. If you get bored at home, this is a sure sign of resistance. Instead of running away, you should stay with it—stay at home and discuss your relationship with your wife, work on it. Every true relationship requires constant *work*.

John: Oh God, work at the office all day, and work on my relationship at nights and weekends. What a life! No wonder I want to get away sometimes—hey, though, wait a minute! Why do you take my boredom at parties as proof of *their* superficiality but my boredom at home as evidence of *my* superficiality? You've got me in total double-bind—damned if I do, and damned if I don't. Well, if that's the name of the game, you can stuff it.

> I'm not *working* on any relationship—besides, even if it were the best in the world, I'd still be wanting to get away sometimes. I *like* lecturing and broadcasting and talking to interesting people—I *enjoy* it—and I'm not ashamed to admit it any more, so get the hell out of my life and stay there.

With the breakthrough came the realization of how he had been conned for years, by topdog's false logic, into refusing invitations he would have enjoyed and feeling guilty about the ones he accepted. (The dreams occurred not only when he actually stayed away from home, but when he had contemplated doing so during the course of the day.) Now John is able to go away or not according to the merits of the situation and his own feelings, instead of being topdogged into a self-torture game. He also realized that he had often accepted an engagement he would have preferred to refuse, simply because he did not relish the thought of staying at home and *working* on his relationship! Women, marriage counselors, and psychotherapists who continually urge men to do this should realize that this kind of nagging produces the absolute reverse of what they want, and I am amazed that they do not know better. However, the habit of moral preaching and nagging is so common that it is hard for the most enlightened of us to break it, especially parents. The other day when I did my usual mealtime nag, telling Fiona to sit up and eat properly, she replied, "Oh dear, Mommy, when you say sit up, it just makes me slip down." I hope I have learned my lesson!

An underdog "vehicle villain" was unmasked in one of our workshops when Dick, a businessman, dreamed of missing his last plane home from Baltimore airport because the car in which an (imaginary) attractive female client was driving him there broke down on the expressway. His marital topdog was not so easily thwarted, however, for when they looked for somewhere to spend the night in the dream, they discovered that all the hotels were full, horribly dirty, or made of glass. It became clear in working on the dream that it was no mere passing fantasy—he had strong sexual needs which were not being met at home but was prevented from doing anything about the problem by a conviction that having made his marriage bed he had to lie on it, even though this involved the Procrustean exercise of cutting himself down to fit. In a dialogue between the two hidden villains of his dream, Dick broke through to a realization that this conviction was not a law written in the sky, as he had been assuming, but an injunction instilled into him by his mother, which had succeeded only in trapping him in the futile self-torture game which the dream so vividly pictorialized.

DREAMS OF FAILING TO GET THROUGH ON THE TELEPHONE

Dreams in which you try to telephone someone only to find that you have forgotten the number or the telephone is out of order may occasionally be warnings to watch out for some breakdown in communications, but they are more likely to be visual puns reflecting a feeling of failure to "get through" to someone in your waking life, or that there is someone whose "number" you can't get however hard you try. Steve, one of our students, had this last problem. When a dream depicted his failure to get his girlfriend's number, he had no difficulty in relating this to the fact that whatever he did to please her in waking life failed. For example, she insisted that he stop behaving like a child, whining and crawling on his belly—yet whenever he did anything independently she was furious at losing control over him. She put him in a typical double-bind situation in which he was damned if he did and damned if he didn't. The dream made Steve realize the futility of trying to "get her number" under these circumstances.

However, the telephone system is a common happy hunting ground for hidden villains and secret saboteurs who frustrate your efforts to "get through" to someone in your life. John had a number of these dreams in which he would find himself wandering around, doing nothing very much, when he would suddenly recall that he hadn't been home for several days. Panic would set in and he would frantically seek a telephone to discover whether I was still waiting for him or not. Invariably the phone was out of order, or he had no dime, or there was no reply at the other end, and he would wake up in an agony of frustration wondering why ever he had allowed himself to get into such a stupid amnesic situation.

At first we suspected that the hidden villain here was the same old home-and-hearth topdog making some kind of comeback, although we were puzzled because many of the dreams occurred when he had not actually been away for a long time and had no plans for a trip in mind. We eventually unmasked a quite different villain when John was able to relate one of these dreams to a tiny "failure of communication" which occurred the previous evening. John had asked in a very disinterested way whether the book I was reading was good or not, and I, sensing that this was a duty question, had replied rather shortly. We thought no more of this until after John's dream, which made him realize that he had indeed been away for some days on an "inner trip" of his own, reading books on subjects that had no interest for me and thinking his own thoughts without sharing them. His duty question had sprung from a guilty feeling that this was terribly unloving behavior, that he ought to be more attentive to me, though he admitted he would have preferred to continue his own trip. So the dream showed him being driven by topdog to rush to the telephone and re-establish communication, only to have the call sabotaged by underdog determined to do his own thing.

The creative resolution came when John stood up for his own need for inner privacy against topdog's demands for constant "loving relationship"—whereupon he was delighted to discover that I valued my own inner privacy too and hated to have it disturbed.

When I recently came across a new method of yoga/meditation and wondered whether or not I should take it up, I asked dream power for advice. I dreamed of trying to answer an advertisement for a new type of yoga called "Petit Point," but I could not manage to hold the telephone number in my head from the paper to the telephone—and when I did, I dialed the wrong number. When I eventually succeeded in getting through, the maid (we don't have one) came into the room with the vacuum cleaner and deafened me so that I was unable to hear a word. In the dream itself I got the distinct impression that I was somehow not "meant" to take this course, and so I hung up. On waking, I conducted a dialogue with the secret saboteur, who thereupon insisted that I had quite enough disciplines in my life at the moment, and that if I were to contemplate taking a course it should be in self-indulgence, sloth, and unmindfulness! Obviously he was an underdog closely in touch with the real needs of my body and mind at the time, protecting me from the bullying voice of my "religious topdog," who constantly nagged me to impose more and more disciplines on myself in the name of total perfection in spiritual growth. He also pointed out that there was "little point" (petit point) in my persisting in this course, since he would sabotage my efforts in waking life if I did, just as he sabotaged my telephone call in the dream—and just as he sabotaged many of my other disciplinary efforts in the past. He advised me to let go, relax, and allow spiritual growth to unfold naturally and spontaneously from within instead of imposing disciplines from without. He reminded me of the words of Al Huang, master of the ancient Chinese art of T'ai

Chi, in his book *Embrace Tiger, Return to Mountain:* "What you need is an acceptance of yourself as you are. You are like a seed. . . . If seeds had goals, there wouldn't be very many flowers."

I have not actually come across a topdog saboteur on the telephone lines, but it is easy to imagine the possibility. For example, a husband longing to communicate with his wife but prevented from doing so by a parental "observe-the-privacy-of-others topdog," or a left-wing politician longing to take up yoga but prevented from doing so by a "reality-oriented topdog," might well find their dream calls sabotaged by their respective topdogs. You cannot know in advance by any theory whether the secret saboteur of your dream is a topdog or an underdog. The only way to find out is to give him a voice, discover his intentions, and then relate the dream to something in your present life. It is useless to interpret the dream as "my regrettable failure of communication" in life generally because this gives you no clue about where exactly you are failing to communicate, so you cannot remedy the situation—quite apart from the fact that "getting through" may actually be the last thing you should do under the circumstances, as I have shown. Each dream must be worked through to a creative resolution, with topdog villains thrown into the garbage pail and underdog villains (or saviors) standing up for their needs openly.

THE VANISHING STAIRWAY, DOOR, ROAD, ETC.

Another very common kind of frustration dream is one in which you are trying to get somewhere you have often been before and find that the road, door, room, building, or whatever has unaccountably vanished. It is hard to imagine how most dreams of this kind could be warnings on a literal level, though in this age of "future shock" it is not impossible. In most cases, however, such dreams indicate that you are in two minds about getting somewhere; and this may be either in the external world or the inner world of your own psyche. For example, when Dina dreamed of being unable to find the road to the church on her wedding day, she had to face the fact that her playful, joyous underdog had no intention of allowing her to marry her rather proper, uptight fiancé. On the other hand, John's final confrontation and workout with his old familiar "home-and-hearth, deep-relationship topdog" came in a dream where the mysterious man who wasn't there had actually made off with the whole staircase, and in this case the problem lay entirely within John's own inner world.

John had asked dream power to tell him whether or not he still had a topdog inside that could seriously damage his life. The dream gave an immediate answer in beautifully simple picture language, by taking him to the Central Office of Information in London (the brain as a storehouse of knowledge?) and bringing him face to face with his first wife. He was then transported to his former home, where he found himself in an attic rewriting St. John's Gospel in modern terms. His wife called him to dinner, and when he tried to make his way back to the attic afterward, he discovered that the staircase had completely disappeared.

Since John had not actually seen his first wife for a long time and had no interest in rewriting St. John's Gospel, he interpreted the dream symbolically in terms of turning over "at the top of his head" (attic) the implications of our dream work for his own theological and religious views—John's gospel. Since the dream showed him going about this very privately in his first wife's house, he clearly felt there was something shameful about revising his lifelong conviction that people should strive to put others first, work on personal relationships, and put down their own needs, even though he had recognized the futility of this "old morality" in his own life from working out his "missing trains" and "telephone" dreams. He was considering this radical change of life philosophy in a thoroughly underdoggy way, as if hoping that topdog would not notice until the job was well and truly done—but topdog had obviously got wind of what was happening and was literally "taking steps" to thwart this new and subversive thinking. In waking life this probably meant crowding John's mind with trivial and domestic events in order to interrupt his train of thought.

In a dialogue with the secret saboteur her identity was quickly revealed as that of his first wife (no great surprise, in view of the first part of the dream and the fact that the staircase vanished in her house). In John's mind she was the chief spokesman for the "Christian" view of life with its catastrophic expectation that the world would immediately collapse into a chaos of ruthless egoism if people did not deny their own needs. And, of course, John's first reaction was to think she was right, but further dialogue soon exposed the lie by recognizing that most of the really destructive behavior in his experience came from people who were trying to be unselfish and do what they thought best for others. The chief character of St. John's Gospel said we should love our neighbors *as ourselves,* presumably because he knew that any effort to deny our needs in serving other people results only in a covert selfishness that is far worse than simple egoism. The new "gospel of John" is based on this principle and asserts that the more we understand and aim sensibly to satisfy our own needs, the more we shall find ourselves moved spontaneously to do the same for our fellows.

DREAMS OF FORGETTING LINES AND INABILITY TO PERFORM

Here, as always, the first thing to ask of this type of dream is whether it could be a simple warning to check out your facts or brush up your performance for a forthcoming event in your life, in which case the dream usually gives some clue as to what this is. Actors and actresses are particularly prone to these dreams just before the opening of a new show, reflecting not only their anxiety about a part not absolutely perfected, but also about the success of the show and their performance generally. The British actress Judi Dench is typical here: she told the *Sunday Times,* "Before *Cabaret* opened, I dreamed a lot about that. There's that famous old actor's dream everyone has—you are being made up for a play and suddenly you're pushed on and you don't know what the play is or what the part is. It's sheer panic, and I had that over and over before *Cabaret.* I woke up feeling sweaty and horrid." She reported reassuringly that the dreams wore off two or three nights after the show opened and both she and it were pronounced a success. Such dreams, based on real apprehension, are perfectly normal anxiety dreams and do not imply any deep-seated conflict unless they continue past the time when there is real cause for concern.

If, however, an actor continues to have anxiety dreams about a part once it is perfected and going well, then he had better ask himself what hidden villain is trying to sabotage his success. It is possible that there is a parental voice inside nagging that he is no good, incompetent, or should be in some other more worthwhile occupation—and this topdog will sabotage both his enjoyment of work in waking life and his performance in dreams. The creative resolution of such a problem lies in standing up for one's own abilities and needs and in chucking topdog out of one's system. In most cases it is self-evident that the actor is perfectly competent and would be quite unsuited to be a missionary, a scientist, or whatever topdog had in mind for him.

I suffered from a version of this problem when I first began to appear on popular television. On the night before my first talk show in England I dreamed that I arrived at the studio to find that we had to model fur coats. As I walked along the catwalk wearing a sealskin coat, I found that my feet would not go where I wanted them to, and I kept stumbling and slipping, much to my distress and humiliation. On waking, I related the dream to my natural anxiety about the talk show but did not understand why the dream depicted my modeling a sealskin coat, as I had not done any modeling for years. Later, the whole dream clicked into place as I recalled my American publisher's request that I go over and be a "performing seal" for them on the publication of *Dream Power*—that is, promote the book on the media.

So here I was in my dream being a performing seal, and not doing at all well at it, so I asked the secret saboteur why he was spoiling my performance. The ensuing dialogue went something like this:

Sab: Performing seal indeed! I never heard such nonsense. You're a serious scientist and psychologist, and you should be back in your office writing learned papers for the benefit of mankind.

Ann: Oh dear. I hate writing learned papers—and I don't believe anyone reads them anyway. I would have thought mankind might benefit more from my talking about dreams on television. I mean, think of the millions who will listen.

Sab: We academics are not in this world to provide knowledge for the millions. They aren't ready for it. The mysteries should be kept within the four academic walls, and anyone who disseminates them to the public is a traitor to our cause.

Ann: (getting angry) Then I'm a traitor because I don't believe that. The day of the guru is over—I keep saying that. We have to give knowledge to the world. People aren't ignorant and stupid any more . . . and besides, I think going on television might be rather fun.

Sab: Fun indeed! We're not in this world to have fun. And you'll get caught up in a trivial ego trip, seduced by the bright lights, and forget about your serious purpose.

Ann: Look, professor, I'm a big girl now. I'll decide what I do and what I don't do. I'm perfectly capable of running my own life, and I'm going to decide what I do with the knowledge I've worked for. There's no law written in the sky that says I have to conform to your outmoded rules and regulations. And I'm as likely to get stuck in an ego trip as I am to grow wings and fly . . . tell you what, though, if I ever do get anywhere near an ego trip, you can come back as my guardian angel and gently remind me of my purpose. In the meantime, get off my back and stop spoiling my life.

This dialogue made me take a close look at that much-abused term "ego trip," and I realized I had used it against myself in the same stereotyped, unthinking way in which it is so often bandied around by those who are caught in the web of their own social roles—whether it be professor, religious teacher, guru, mystic, or whatever—that they cannot see beyond them. My "professor within" was on a far greater ego trip than underdog rebel, for the simple reason that he was unable to let go his role and expand into something more. This basically is what expansion of consciousness is all about—the opening up of the personality to new possibilities and modes of being, particularly to those aspects of ourselves we despise and reject, which is the only way to real ego reduction.

This is a question on which people who take up religious and spiritual disciplines are especially prone to get misled. As the wisest religious leaders have always known, anyone who deplores egoism or posturing in others is automatically betraying the fact that he is himself on the biggest of all ego trips inside—the ego trip of demonstrating how perfectly "faceless" he is. This is beautifully put in a Jewish story, told by Perls in his book *Gestalt Therapy Verbatim,* about a rabbi in a synagogue who started to protest his nothingness before God. The cantor took up the theme, and then a little tailor in the congregation began to do so, whereupon the rabbi turned indignantly to the cantor and asked, "Who does he think he is to think he's nothing?"

I wish I could report that my ego-ridden academic topdog stopped bugging me immediately after our dialogue, but so deep was his catastrophic expectation that I was selling out that he spoiled my enjoyment of several good shows. But I persevered, and when he was convinced that I was not doomed to perdition, he gradually let me off the hook and even joins in the fun himself now—never forgetting his very serious purpose, of course!

Not so with Jane, for the secret saboteur who spoils her performances is an underdog—and underdog has basic needs that must be met. Jane, a housewife with four children, is also a professional pianist who is so overcome with anxiety before performances that her subsequent success is barely worth the strain she endures. Her dreams reflect a deep conflict; in one typical frustration dream she is in church, the minister calls the hymn number, she tries to play but can't, and the congregation sings without her. In another, her children run around making a noise so that her playing cannot be heard; and in yet another, she is playing in a restaurant only to discover that her music is being distorted by the amplifying system. Complaining to the manager, who says it can't be helped, she returns to the piano to find that the back leg has fallen off. When we first worked with Jane we suspected the presence of a "domestic topdog" derived from her mother, nagging her that any woman worthy of the name would devote herself totally to the family, but many hours of work spread over a long period of time failed to resolve the conflict. Despite the fact that Jane told topdog to get lost time and again, the dreams continued and her anxiety attacks about her playing and life in general became worse.

The clue came in a dream in which a black man brought a crippled girl to me to be healed, and Jane was very concerned about whether or not I could do so. She helped me put a robe over the brace on the child's leg but never discovered whether healing took place or not. When Jane gave the child a voice, she said, "I'm a crippled child, and you can't expect a crippled child to play the piano." We then realized that the secret saboteur was not a nagging domestic topdog but a crippled underdog who was quite determined not to perform under duress. We also realized that unless Jane resolved the conflict soon, this underdog would not hesitate to cripple her in waking life as well in order to get his own way.

Further work revealed that both parents had encouraged her to be their "big, strong girl," and her mother in particular had insisted that she develop her talents and play the piano —perfectly, of course, just as she insisted that Jane perform perfectly at everything she undertook in life generally. Both parents were delighted when Jane became a success, as this raised them socially, and although Jane enjoyed it too, underdog was secretly furious at being forced beyond his pace and used as a status symbol for the glory of the parents. He quickly sabotaged Jane's career by leading her into marriage at an early age and producing four children in quick succession, so that for many years she was too busy to perform. When they started growing up, however, the mother within started nagging that Jane should play again—perfectly, of course—despite the difficulties of coping with a large family. Jane did play publicly from time to time, at enormous cost to herself, but the voice of approval from this inner topdog made her feel so good that she pressed on with the good work of "self-improvement." Eventually, she became so unhappy, bitchy, and ill that she sought therapeutic help.

Her problem, briefly stated, was the old familiar bogey of perfectionism—not only in her playing, but in everything she did. Her whole life was geared to please mama, irrespective of what the inner Jane wanted. She had bought the catastrophic expectation that to be imperfect in anything she did automatically dubbed her a failure and a nonentity. Jane's situation reminds me so much of Stuart Miller's predicament, described in his book *Hot Springs,* that I will quote from it briefly here —and I feel it will reach many hearts. Stuart and a party of Residents from the Esalen Institute hiked to Tassajara, a Zen monastery in the hills of Big Sur, California. The rest of the party determined to live like the monks for a few days, undertaking their tasks and disciplines, but Stuart, realizing the futility of the operation, kept himself out of it:

> I dwelt happily with Clarissa fucking and drinking wine. I was glad to have survived the hike, glad to be in such an exotic place, glad that Bo had failed, glad that the group was being so absurd. When at five-thirty A.M., the great ceremonial drums and gongs began sounding monks and my Residents to morning meditation, the dogs running with the boys and girls and men, I sneered as I turned over in my sleep at the pathetic attempt of worldly failures to try and imitate such serious people.
>
> It never occurred to me that two days of meditation might be the beginning of something, maybe not a life of Zen meditation, but something. I had learned how to avoid both failure and experience. When I was fifteen the creative-writing teacher

> told me that my first short stories were lousy and I decided not to be a writer. No new things for me, thank you. Excellence, excellence. If you can't be excellent, don't try. Die, you incompetent mediocre son of a bitch, only test pilots in this man's air force!

While Stuart chose to be nothing rather than risk failure, Jane chose to try to be a test pilot, at the cost of psychological suicide—the murder of the child within who asked only to be itself and develop its talents and attributes in its own way, at its own pace, without force or strain. Jane has the catastrophic expectation that if she accepts this child as part of herself, she will automatically become a failure, a do-nothing housewife, and as she wrote in a letter to me, "I still can't believe that either you or I would be happy with a do-nothing, non-playing Jane." I pointed out that she had already been forced into a pretty do-nothing situation by constant anxiety, depression, and illness and was neither a good player nor good housewife at the moment. She was, in every sense of the word, at impasse.

Getting through the impasse means taking a risk of losing one's precious self-image and disappointing the parents within (and maybe the real parents too) who expect total perfection. The amazing thing, of course, is that once through the impasse, you find you have lost nothing at all—in fact, you find *yourself.* Jane is still working through to this point, and in one of her dreams she found herself watching a small, dark-skinned girl playing the piano with unusual gift and ease and was amazed at her ability. But the child was plain and homely, almost black (a sure indication of a shadow figure), and Jane thought she would not have much chance as an adult performer unless she became much prettier and more sophisticated. The child is Jane's own—that part of herself she locked away in childhood because it made mistakes and did not live up to topdog's expectations of perfection. The more

Jane strove toward a bright and beautiful false self, the more crippled the dark child became, with the result that Jane could play only with difficulty and anxiety—for after all, you can't expect an imprisoned crippled child to play the piano. The resolution lies in defying topdog and bringing out the small dark child, accepting it as part of the self with all its childish imperfections, and allowing it to grow to maturity in its own special way.

Jack Downing summarizes the position in his book *Dreams and Nightmares:*

> Loving myself as I was as a child is absolutely the crossroads of personality development. If I can't find love, compassion, acceptance, understanding for that early self, then I sentence myself to travel a lonely road, destination perpetual loneliness, emptiness. We are so limited in our powers as children, we are so totally at the shifting mercies of forces entirely beyond our control! We take the words we hear so absolutely literally! As adults, we often wrap ourselves in the symbols of power, money, position, connections, and despise those persons without power; we also despise ourselves when powerless, when helpless children, we despise our inner self, our child, still present, still hungry, still suffering, still afraid. "Unless you become as a little child, you cannot enter the Kingdom of Heaven." That "little child" is *you,* your own inner self, your child.

DREAMS OF BEING UNPREPARED AND ARRIVING LATE

Dreams of being unprepared and arriving late for some function have the same inner dynamics as dreams of forgetting lines and of being unable to perform. If you have one, first try to identify the event or function to which the dream refers by examining the dream characters and setting and then asking yourself if you are really prepared for it in waking life. Or the dream may be saying that you are just "not prepared" to do something—some task, role, or action—in your waking life. If you really are prepared for the event or to take some action in your life, then you must ask yourself what hidden villain is sabotaging your efforts and why.

A typical villainous maneuver in these dreams is the removal of one's clothes or something essential for the event in question. For example, Elaine dreamed that her mother ordered her to go out into a storm with a friend and she agreed, feeling that it would be good for her, but she was unable to find her boots and knew her mother would never allow her out without them. Since she has not seen her mother for a long time, Elaine realized that the dream mother was her own inner topdog whose nagging voice had been evoked by a crisis that had "blown up" at work. Her mother topdog was urging her to face it openly, honestly, and courageously instead of dodging it like a sniveling coward. But underdog had no intention of facing the storm and arranged to get his own way without losing topdog's approval, by hiding the boots. The dream warned Elaine that if she decided to "face the storm" in waking life, her underdog would be looking around for a way to sabotage her good resolution, perhaps by illness or some other means.

An underdog saboteur was also at work in the life of a newly ordained minister who called in when I appeared on a radio talk show in Atlanta, Georgia, with a dream of being unable to find his Bible just before an important service. Since he could not resonate to being literally unprepared, we probed deeper and soon found that he was having grave doubts about his vocation, which he had taken up because it had always been his parents' ideal for him.

On the other hand, it turned out to be a topdog who tried

to sabotage Elizabeth's attempts to remarry after several years of loneliness following the death of her husband. In a typical series of frustration dreams she would set out to meet her fiancé only to find that there was no gas in the tank or that she had left her pocketbook at home, or that he had left home by the time she got there. Well-meaning friends with a smattering of dream knowledge had suggested that she did not really want to marry this man, but she could not resonate to this and asked me for help. Dialogue with the secret saboteur revealed his identity as her late husband, who upbraided her for daring to consider remarriage and thereby expunge all memory of him after thirty-three years of life together. Elizabeth recognized this voice as her own "faithful-forever topdog," derived from her husband, and was able to resolve the conflict by catching him out in the lie that she was not a cheap, unfaithful, superficial person for wanting to remarry, nor would this diminish the experience of their own life together. On the contrary, she insisted, their experience of marriage had been so good that she was prepared to try it again.

THE TOILET THIEF

I cannot close this chapter without mentioning one of the commonest frustration dreams of all—that of bursting to go to the lavatory and being unable to find one, or finding one and being unable to use it for some reason. Sometimes we awaken from the dreams with a full bladder or bowels, but not always —and in these last cases, the secret toilet thief almost always turns out to be a topdog who will not allow us to release our "shitty" (e)motions, such as resentment, anger, fear, and so on. Indeed, he would prefer that such emotions did not exist at all, for he is determined to remove all outlets for them.

These dreams can usually be traced back to a point in the day when we withheld the expression of these emotions at topdog's command, and so we frantically seek relief in dreams. But topdog will not even allow us to "dream" of doing so. Almost always a self-image of perfection is at stake, and topdog sees that we live up to it. What he fails to understand is that you cannot block an emotion once it has started to flow through the body without incurring some physical or psychological damage—for emotion is the *physical* reaction to a thought or perception. What normally happens is an inappropriate explosion in an inappropriate situation, which causes far more hurt and distress to ourselves and others than a legitimate expression of feeling would have done in the first instance. Down with toilet thieves!

THE HIDDEN HERO

Hidden heroes who mysteriously avert disaster at the last moment in dreams are not nearly so common as hidden villains, but most of us have experienced them from time to time —for example, the hidden hand that steers the car to safety as it plunges down a precipice, or the mystery voice that whispers answers in your ear in an examination you feel you are bound to fail. These seem to me to represent a principle quite different from either topdog or underdog—the principle of our deepest health or strength which makes it possible for us to break out of our self-torture games. Jung called it the Transcendent Function, and in many dreams it is not invisible but finds embodiment in some character or symbol that carries a great deal of emotional power.

I shall return to this very important subject in Chapter 13, but I want to emphasize here that the discovery of this unsuspected resource of health and strength inside us depends, in my experience, on our readiness in waking life to seek wholeness or fulfillment in terms of what we actually are and not in terms of what some authority, outer or inner, tells us we should be. If we try to resist the process of being ourselves, the hound of heaven will be after us to return us to our center which contains everything we need for our health and growth.

«11»

NIGHTMARES

God turne us every dreme to goode!
—Geoffrey Chaucer

"Nightmares, according to recent evidence, are horses of at least two very different colors," writes Ernest Hartmann, editor of the symposium-volume *Sleep and Dreaming,* produced in 1970 for the International Psychiatry Clinics. "A heavy, shapeless black beast crushing the sleeper's chest as he awakens in terror, and a more ordinary reddish mare galloping off with the sleeper on a frightening and yet relatively familiar dream journey." The first type is known as the "incubus attack," which usually occurs during the first two non-REM periods of deep sleep and is accompanied by a single terrifying scene of falling, choking, being crushed, or whatever, and a sense of impending doom. The second, less severe, type of nightmare is the familiar drama of being attacked, pursued, drowned, and so on as the culmination of a long, ongoing dream; this usually occurs toward the end of one of the last REM periods of the night.

NIGHT TERRORS—AND OTHER DEMONS OF THE DARK

The incubus attack includes the night terror *(pavor nocturnus)* of children (and some adults), which often culminates in a bloodcurdling scream, palpitations, choking, and a feeling of paralysis. It is believed to be physical in origin, a disturbance of the arousal or waking-up process for which there is yet no adequate explanation. If you or your children suffer from such night terrors, you should not worry about them too much, for while psychology and medicine have no idea yet what causes them, it seems clear that they are not likely to kill you. I personally have no experience of incubus attacks, nor have the subjects who contributed to my research, so I can only suppose that they are relatively rare. I mention them here in case you or your children have experienced them and wonder how to deal with them. My advice is to comfort a child until it is ready to return to sleep, explaining that there is nothing to worry about, and you will find that it is usually forgotten by morning. Most children grow out of these attacks eventually with no apparent harm done. In the case of adults the sensible course of action is essentially the same—you can be fairly confident that the attack will not recur if you go back to sleep, nor are you any more likely than usual to have bad dreams on the same night. There is little point in trying to interpret incubus attacks as signs of psychological trouble since they have so little dream content; the most you can do in this direction is to look at your ordinary dreams around the same time to see if *they* show any evidence of something wrong in your life.

REM-period nightmares, however, usually occur toward the end of a long REM period in the latter half of the night, almost as though the dream were building up to its climax of terror. Most of us have our own favorite breed of "mare" (which actually means demon, not horse)—being chased, attacked, burgled, executed, imprisoned, drowned, or whatever —and when it recurs in a dream we are upset but not altogether surprised because we have undertaken this particular trip many times before. My own work on dreams indicates that a recurring nightmare invariably indicates a deep split within the personality, and the more severe your nightmares, the more severe the split. My experience has been, however, that intensive work with nightmares along the lines I shall describe results not only in their disappearance but also in the healing of the split, with the release of an enormous amount of energy for constructive living.

For many years I was puzzled by conflicting views of nightmares (which I define as dreams bad enough to awaken or worry us) by experts of the various schools of psychology. For example, Calvin Hall writes in his book *The Meaning of Dreams:*

> Many dreams are unpleasant, even terrifying as in the case of nightmares. The dreamer is being chased by a man with a knife, a burglar is entering his room, a lion is about to spring on him, the house is on fire, he is drowning or he is arrested and put in prison. These are all punishment dreams. For what reason is the dreamer being punished? Because he has violated one of the commandments of his conscience. He has rebelled against authority, or he has gratified a forbidden wish, or he has committed a misdeed. The nightmare is the price he pays for doing something wrong. Such dreams provide us with information about the dreamer's conceptions of the penalties that will be inflicted upon him should he disregard his conscience and yield to temptation.

Examining my own dreams, I found many nightmares that were almost certainly self-punishment dreams for some offense committed against an external authority or inner ideal during the course of the day, and these usually took the form of imprisonment or execution. In my case the major sin was usually "giving in" to some emotion—anger, fear, joy, grief, jealousy—and allowing myself to be swayed by it. On the other hand, there were nightmares in which I was threatened by a tidal wave or wild animal, and I was usually able to trace these to the suppressive control of emotion during the day, in the sense of trying to rise above it, in keeping with my self-image of a cool, detached, spiritually evolved person. I seemed to be caught in a double-bind in which both the expression and suppression of feeling resulted in a nightmare.

It seemed, judging from my reading of Jung, that my tidal waves and animal pursuers were much more like what he called the "shadow," that part of myself I considered unacceptable and wanted to rise above. But suppressing and "controlling" my feelings and "animal needs" during the day merely caused them to gain energy surreptitiously and they would hound me at night in search of acceptance and integration. John Sanford summarizes the Jungian view very simply in his book *Dreams: God's Forgotten Language,* as follows:

> The persona is the mask which we wear before the world; it is the front we put on. Partly the persona is a matter of social necessity, a form of adaption to the society in which we live. But too often we come to identify ourselves with our persona. We think we are the person whom we would like to appear to be. The shadow stands in direct contrast to this persona. The more we try to appear all goodness and kindness, the more brutal our shadow appears in contrast; the more we try to be all strength and courage, the more we are followed by a shadow of weakness.

So it seemed that my nightmare figures could be either conscience in search of retribution or the shadow in search of acceptance. Neither the Freudian nor Jungian theories offered an adequate model of nightmares, and it was only when I came across Gestalt therapy with its topdogs and underdogs that my whole line of thinking clicked into place. It quickly became clear that a nightmare was merely a severe case of the self-torture game actually manifest in the dream itself, with one clown being hounded by the other, according to which one had offended the other during the course of the day. A "self-punishment" nightmare occurs whenever underdog has defied topdog during the day, in which case topdog hounds him at night in order to regain control over him. On the other hand, whenever topdog has denied underdog a really basic need during the day, underdog, driven to the end of his tether, comes out in a thoroughly nasty way at night. In either case, it is a fight to the death, which accounts for the feeling of dread and terror in the nightmare itself.

I am not saying that this model is applicable to all nightmares but it is a useful one, particularly in the case of nightmares in which you are pursued and attacked; and it should always be tried for size when coming to work on a dream. It is crucial to know whether your nightmare pursuer is a topdog or underdog, because you have to learn to deal with them in quite different ways. The Jungians who advise us to "agree with thine adversary" are quite correct if we are dealing with an underdog or shadow because you cannot run away from your own shadow. The creative resolution lies in accepting and integrating him into your personality, whereupon he loses his darkness and terror and adds to your stature as a "whole" person. It is the old story of Beauty and the Beast in which the beast turns into a handsome prince when the princess is prepared to accept and love him for what he is. But if your dream adversary is a topdog, this is the last thing you should do. He too is a part of you, but you *cannot* agree with him, for his aim is to rule you completely, and the more you try to agree with him, the more his demands increase—in contrast to underdog, whose raging character disappears as soon as you pay attention to his needs. Topdog must be confronted, overcome, stripped of his power, and given the status of servant. Needless to say, he doesn't like this at all and will try every argument under the sun to retain control of the personality.

The following examples will illustrate my case and demonstrate the measures you can take if you really want to rid yourself of nightmares and heal the split within yourself. (To save space in the accounts themselves, I will mention here that in all cases, possible interpretations on an objective level, both literal and symbolic, were eliminated before moving on to the deep subjective level. Many nonrecurring nightmares, particularly those of small children, are pictorial reflections of threatening people and events in the dreamers' lives at the time of the dream, and to interpret them as signs of a split in the personality would be to miss out on something very wrong in the immediate environment.)

DREAMS OF PURSUIT, ATTACK, VIOLENCE, AND INTRUSION

Scott dreamed:

> I am standing on a crowded beach with a group of friends from varied previous chapters in my life, two of whom are Milt and Len. I jokingly chide Milt about something and lightheartedly run a few steps to avoid the imaginary impending recriminations. Suddenly I know I really do have to escape—and as I run they fire a bullet smash into my face. They chase me into a neighbor's house and corner me in the living room. Len orders me to sit down on the couch and threatens me with his gun. He tells me to relax and just accept the fact that I am going to die. I accept this, calm down, and try to let go my need to live.
>
> Then suddenly a childhood memory flashes into my mind. I used to read books on war captives, and I had resolved that if I were ever at the mercy of barbaric captors and faced the prospect of death, I would not passively take it—I would either turn on my captors and go down fighting, or break free and run, knowing that my attempt was futile and that I would be

> gunned down. So I jumped up from the couch and dove through a window. Two shots cracked out and as I heard them punch into my shoulder, I awoke. My arms were drawn under my chest, rigidly contracted, and my head straight out in battering ram position. Every muscle in my body was on the verge of bursting from the intensity of the dream. When I realized that it had all been a dream, I sighed with relief, rolled over, and relaxed.

Scott worked on this dream with two other students without any personal help from me, and this is his report of what happened:

> I had this dream about a month after starting college, and I was in a very bad state. I was frightened by everyone and simultaneously in need of *any*one, for I felt totally isolated. Compounding this was my inability to show feelings, which I constantly attempted to keep within the bounds imposed by the credos I live by: "Everyone is likable" and "There is never any need to get uptight."
>
> The dream was probably sparked by the fact that Len (who had been stealing his way through the U.S. when I picked him up on the road and took him home about a week before the dream) had become very fond of one of my records, which happened to be missing the day he decided to pack up and leave. I asked him offhandedly if he knew the whereabouts. He'd lent it to a mutual friend, he said, who had left town with his cabin locked. I was angry, frustrated, yet helpless. I didn't want to demand the album, for though I strongly doubted it, he might have been speaking the truth. So I decided to live up to my credo, pulling my anger and frustration back inside the tranquil bounds of "no need to get uptight."

But it was too late. Scott had already sinned against the ideal of perfect love and trust demanded by topdog, in allowing himself to *feel* resentful, angry, suspicious, and unloving toward his friend and in daring even to ask about his record.

So out comes topdog to hound him in the dream, to punish him for his transgression against perfection. Scott continued:

> We arrived at the emotional impasse in the group when I had a dialogue with Len. Len was right, and I agreed to do better next time. I promised never to distrust him or others again. I asked him to forgive me. Then Sheri suggested that I ask Len if the death sentence was not rather severe, and the subsequent dialogue went something like this:
>
> *Scott:* Why are you killing me? I haven't done anything so terribly wrong. Leave me alone.
>
> *Len:* I've told you what you've done wrong—and anyway, I *want* to kill you.
>
> *Scott:* But why? I didn't do anything.
>
> *Len:* That's just it. You never do—and I'm going to make you. You're always sniveling around, you little creep (leaps up and reaches for gun).
>
> *Scott:* Please don't—wait. Let's talk this out. We don't need to use violence to settle this.
>
> *Len:* *I'm* settling our differences—I'm the boss here—and I'm going to do it now. Die, you whining little dog. Jump. Do something.
>
> At this point, he was lunging at me. I grabbed for his knife and kicked him as hard as I could, the excitement uncharacteristically exploding in me, a powerful mixture of anger and strength. Topdog had finally forced sniveling Scott into action.
>
> The episode didn't resolve the conflict completely: I have it with me yet, but I'm becoming more tolerant of parts of myself I don't like and also more willing to take risks with others.

This is a beautiful example of the way topdog puts underdog in a total double-bind, by first insisting that Scott live up to his expectations of the meek, docile, passive individual who accepts kicks and hurts without complaint, and then chastising him for being a doormat. It was the latter that forced Scott to break through the impasse into action, for he realized that there was no way he could win this battle. There is only one thing to do with a topdog like this—and that is to throw him into the garbage pail at the first opportunity. If he chooses to return in the form of a servant, all well and good—but it takes some time before he is willing to relinquish his control. In the meantime it is up to Scott to be consciously on the alert for comebacks similar to the Len episode, and to catch topdog and squash him before he does more harm.

While Scott did exactly the right thing in turning and fighting his adversary in waking fantasy, it would have been quite wrong for Carol to do the same with hers. She approached us at a dream conference to ask help on a dream with a nightmarish ending that continued to trouble her. In the dream she was making her way to the rest room, when a large black man pushed his way in front of her, stood in front of the washbasin, glowered at her threateningly, and refused to move.

When we asked for the day's events, it emerged that she had received a colonic irrigation (enema) as part of the physical cleansing process accompanying a higher spiritual endeavor. This explained why the dream took place in a bathroom—and it seemed clear that the black man wanted to prevent the purification process. As she could think of no reason why "he" should want to do this, we asked her to give him a name and talk to him. George seemed to fit, and the ensuing dialogue went something like this:

Carol: Why are you following me around and pushing in like this? Get out.

George: No, I won't. I'm sick and tired of always being pushed out. You always manage to get in first, and it's my turn now.

Carol: Your turn for what? Get out and mind your own business.

George: I'm tired of minding my own business and being

pushed around. You never listen to me, never pay any attention to me.

Carol: Why should I? I'm trying to be clean and pure, and you're so ugly, black, and dirty. You're spoiling it all for me, following me around like a dog.

George: That's just the point. You're so concerned to be pure that you push me away. And I'm not dirty at all—I'm just black. Well, I'm at the end of my tether now, and if you push me away once more, I might get violent. I shall deliberately sabotage all your efforts at purification, and I shall impinge on your prayers and meditations and dreams.

Carol: But I don't know what you want.

George: I want to enjoy life. I want to sing, and laugh and play and dance.

At this point in the dialogue Carol started to cry, and through her tears she kept repeating the words "I want to dance, I want to dance." It was clear that in the pursuit of "higher things" she was repudiating her shadow (often depicted by blacks in dreams because they are dark, oppressed, and mythically "animal" in nature), trying to have day without night, light without darkness, purity without blemish—Carol without George! Like the blacks in society, the shadow becomes threatening and dangerous only when oppressed, its violence stemming from impotence in the face of insuperable odds. She realized that for a long time she had wanted to laugh, sing, play, and dance, to move her body freely and joyously with the abandon she had seen in black dancers. Eventually, she got up and danced with us, and she was still dancing when we left her—a thousand times more truly "spiritual" than before. Irrigation and cleansing had to come from within in the form of tears, washing away her tightness and constriction, not from without in the form of enemas. It seemed to us, and I think it felt to Carol, like a mini-satori, an explosion through tears into joy.

John Sanford, continuing his description of the shadow, writes:

> The more unconscious we are of our shadow, the more autonomous and undifferentiated is his behavior within us. Then he is likely to erupt with violently disturbing manifestations, or he can place himself—an unseen entity—between our intentions and our achievements. But when the shadow is given conscious recognition we will find that he has much of positive value to add to our personality, often giving us that earthiness, or instinctuality, or zest for living, or humility, which is exactly what we need to complement our personality and become a complete human being. It is a strange thing about the shadow: for all of his darkness, he seems close to God.

ANIMAL NIGHTMARES

The shadow often takes the form of an animal in dreams, for the whole of Western civilization has been built on the attempt to subdue animal nature, both inside and outside ourselves. From Plato on, Western education has been based largely on the notion that the mind has to control the body like a rider controlling a horse and that if control is ever relaxed, the horse will not merely run amok, but will probably degenerate into a ravening wolf, a wallowing pig, or a lazy slug. Only in recent times have we begun to realize that this is an altogether misleading model of human personality and a libelous view of the animal kingdom into the bargain. In natural conditions wolves are far more loving than destructive, pigs are not particularly dirty or greedy, and slugs are not lazy—while as far as human nature is concerned, mind and spirit can no more control the energies of the body than a flower can control the plant out of which it grows. They must grow and work together in harmony, understanding that the health of one is dependent on the health of the other.

When fierce animals like lions and tigers pursue us in dreams, it is usually a sign that we are trying to be "civilized" at the expense of our inherent natural dignity and power. Psychologists normally call these energies in us "aggression," but this term has overtones of destructiveness which does us (and the lions and tigers) less than justice. The very first time Fiona saw a tiger in the London zoo, she exclaimed, "Oh, look at the noble prince!," instinctively recognizing a kind of wild, very sexual grace which human beings repress in themselves (witness the bloodless eunuch princes of fairy tales!). The wildness becomes aggressive only if the creature needs to fight for survival and freedom—which is the point at which our own "animal nature" turns on us and pursues us in dreams, because the life-style we have been trying to create for ourselves lacks something absolutely vital to our well-being.

For example, as I come to the close of this book, my dream tiger is becoming more and more impatient to leave the gentle climes of Florida and seek out the spaces and wildernesses in other parts of the world. Only last night, he got out of his pit in the yard (that's where I was keeping him, and it seemed quite logical in the dream) and chased us all into the house, and on another occasion when I was flying in a dream, I saw him down below and waved to him, whereupon he growled threateningly. My animal nature—my need for "wildness"—is becoming restless and impatient because it has been denied expression for so long. It is no life for a fine, fearless creature of the jungle to sit all day at a typewriter, having no wilderness in which to roam, putting on weight for lack of exercise, and getting rusty in all joints except the fingers—and a bit of daily yoga is no substitute! I have promised him a change of scene very soon, though until the book is done, I have no option but to "rise above him." I am reminded of my recurring nightmares during pregnancy, when I was pursued by wild cats, dogs, and wolves, which reflected the way in which I was putting down my body's wish to become really involved with maternity. I was, in fact, trying to rise above all such feminine nonsense by working right up to the last minute and treating the birth as a total sideline to my Ph.D. My catastrophic expectation was that if I allowed my hormones an inch, they would promptly turn me into a maternal cabbage.

In our workshops we are particularly struck by the fact that many pale, sickly-looking students who are striving to live up to the counterculture's standards of universal gentleness and calm have more than their fair share of animal nightmares, which is almost always an indication that they are denying their animal energies in a most dangerous way. It seems to me that the counterculture must devise some constructive way of allowing expression for the body's energies other than flute playing, listening to music, writing poetry, or making pottery. When we were told that the lion shall lie down with the lamb, it surely does not mean that the lion should become a lamb. Jesus, who has been called the Lamb of God, did not hesitate to roar like a lion on occasions (usually at topdogs), and it seems to me that Blake expresses the spirit of psychological truth when he writes in *The Marriage of Heaven and Hell:*

The pride of the peacock is the glory of God.
The lust of the goat is the bounty of God.
The wrath of the lion is the wisdom of God. . . .

The roaring of lions, the howling of wolves, the raging of the stormy sea, and the destructive sword, are portions of eternity, too great for the eye of man. . . .

The eagle never lost so much time as when he submitted to learn of the crow.

Another very common "body underdog" that causes a great deal of trouble in our society is the one that makes people overeat and overdrink. Sara, whose dream glossary appears in the Appendix, is a perfect example of the futility of trying to deal with such an underdog by repressive disciplines, and as you will see, she cured her drinking by trusting her underdog's own innate nature instead of flogging him to death—which always resulted in excess. The body has its own controls, and if we overindulge, then the trouble lies in the mind and life-style, except in rare cases when the glands are at fault. This is not to suggest that you should never diet or discipline in moderation—only that if you fail time and time again, then you had better take a different approach and ask the body what it *really* wants, for this is a sure sign that you are denying it something essential.

The following dream of Gladys Davis Turner, secretary of the late Edgar Cayce, illustrates her balanced approach to the problem of diet and her heart's response to it. Here it is in her own words:

> First this background: for several years I have avoided chocolate because I am somewhat allergic to it, having overindulged in my younger days. On Sat. 9/15/73, I left the A.R.E. Board Meeting at 7.00 P.M. where I had acted as secretary since Fri. at 4.00 P.M. with the exception of eating and sleeping. We had continued the meeting without dinner so that we would not have to return that night. On the way home, I stopped at the grocery store to get a few things, thinking I would also pick up something for a snack, since I didn't want a full meal, having done nothing but eat and sit for two days. Going up and down the aisles, I found myself at the candy counter, and my eyes lit on a box of milk chocolate almonds, ¼ or maybe ½ lb. I thought, "I haven't had any chocolate in a long time. The chocolate is nourishing and I know the almonds are good for me. Maybe it won't hurt me to eat a little like this occasionally,

maybe once or twice a year." So I bought it and on the way home in the car I kept eating it until I had finished the box. It was *so good* and satisfying. I thought it was even worth having a headache. After working round the house a while, I drank a glass of milk and went to bed. No bad effects next morning, and I remembered this dream:

I was working in my home, doing this and that, very busy. Off my nice, big kitchen to the right was a big family-type room, like a closed-in sun room or garage, but nice and orderly, comfortable. I kept thinking, "I've got to stop now and feed that pig." Still, I went on finishing up this job and that, until finally I said to myself, "It's been *days* since I fed that pig—I've just got to stop and feed him!" I went into the sun room where the pig was kept in a big square box like a hat box but made out of wood; not a fence but a covered box. I opened the top and there looking up at me was a beautiful little white pig (snow white), with blue eyes, who smiled at me!

Since Gladys's own eyes are blue, she had no difficulty in seeing the dream's meaning and could appreciate its humor in parodying her catastrophic expectation that her decision to buy the chocolates, taken entirely on the merits of the situation, would be the first step on the slippery slope to the pigsty! Had she continued to deny her body's perfectly reasonable needs, she might have found a wild boar breaking out of the sun room. Gladys's dream also illustrates another psychological fact—that when we do pluck up courage to defy topdog and pay attention to the needs of underdog, he immediately becomes our friend, even though we may have neglected him for a long time. Underdog is a forgiving creature uninterested in revenge. As one of John's dream tigers purred as he tentatively stroked it with his foot, "It takes so little love."

DREAMS OF TIDAL WAVES AND DROWNING

Jeff, one of our first-year college students, dreamed:

> I am in the basement of my house when the phone rings. It is a friend of mine, Bill, asking me to come quickly to his house. I do so and ask him what's wrong. It seems he needs some help taking off his shirt or jacket. I do so, but then I feel something else is wrong, and I look round the house to discover what it is. The first place we go to is the attic. There are small statues lying around, and it is fairly well lighted. There are five light switches on the left of the doorway which turn on. After glancing around quickly, I find nothing of interest and turn off the switches, but not all the lights go out. No matter how hard I try, I can't get them all off. One in particular, a blue light shining on a blue statue, just won't go out.
>
> I turn around and see my friend and brother walking up the steps toward me. My brother is very annoyed that I am in the attic fiddling around with the lights. I explain that they won't go out, and demonstrate. . . . looking up, I see a stained glass window in the ceiling and hear water splashing against it. The next instant the glass breaks and water comes crashing down on everything. My friend and brother run down the stairs with me arriving downstairs a few minutes later. Water has come right through the house as if a great flood had hit it, and my brother orders me to one side of the room saying that the house is leaning and will tip over if my weight doesn't offset it. I do so with great resentment at being ordered around. When I get to the side of the room, I see a trapdoor leading to the basement, which is rapidly filling with water. At the bottom stands my twelve-year-old sister, and I am frantically trying to save her from drowning as I wake up.

Jeff worked on this dream alone as part of his end-of-term paper, and he continued with his interpretation:

The attic is a church with the statues and all the lights and—what really makes it clear to me—when I shut off all the lights, the lights on the blue statue remained on. I'm a Catholic, and I don't know if you have ever seen a Catholic church with all the lights out, but I've stopped in our church at night when no lights were on and everything would be dark, except for the blue statue in the front which, as in many churches, is lit up by candles. The stained glass windows also leave me with the same conception.

I asked myself, "Why did I go up to the attic?" and the reply was something like this: something in my life at college is wrong and in trying to find out what, I first examine the church —that is, the teachings of my church. All throughout my childhood and high school years, my life at home has been influenced by my parents' morals. I never considered not going to church because I knew my parents expected it of me. Upon arriving at college, however, my morals have changed quite a bit. I am doing quite a bit of thinking for myself, and have done things I would never have considered doing at home. Every emotional conflict I've had since I arrived has been based on moral issues, and it has been using up a lot of my time and energy. Maybe because I was afraid of what I might think of myself, I kind of locked up these concerns in the back of my mind and went on with life assuming them forgotten. My dreams seemed to say different, however.

My friend Bill and I are the only ones who have left the north (place of cold, Puritan life) for the south (warmth, passion, sex), and I am helping him strip away his moral prohibitions (taking off his clothes) which shows that my heart is basically on the side of breaking out of the old restrictions. I succeed—but then the wrath of God descends. I remembered Calvin Hall mentioning "basement" in his book *The Meaning of Dreams* and so I looked it up—a place where "base" impulses are kept locked up—and the church (the attic) is where my religious morals are kept. I associate the water pouring down with "then there came the Great Flood drowning all the

> sinners of the land. . . ." that is, I feel the wrath of God descending on me for not sticking to my moral code. I choose the flood of water, just as the Old Testament used it to portray God's wrath!

Jeff conducted a dialogue with his dream brother, who tells him that he is "rocking the boat" by his loose morals, and he must get himself back on the side of right or the whole house will topple. This, of course, is the typical topdog lie—but Jeff buys it in the dream and does as he is told, with great resentment at being bullied. Another dialogue, with the sister, went like this:

> *Jeff:* Why are you in the basement? It's very dangerous.
> *Sister:* It's your dream. You put me there and it's your fault if anything happens to me. See what you've done fiddling with those lights. You should have left them alone.

Jeff concluded his paper as follows:

> She made me feel guilty about the turmoil I'd caused. In the same way, I would feel guilty for the trouble I will have caused my parents if they find out how my beliefs now differ from their morals. The dream enabled me to sit down and have a talk with myself: to resolve the conflict, and no longer to expend the energy necessary to continue to keep the problem out of my conscious mind. Dreams to me have now seemed to take on a more meaningful form. They seem to be beautiful creations written and produced by me, which have so much to say about me that I regret so many have passed me by.

Jeff does not say exactly how his conversation with himself went, but I hope it was something like this:

> *New Jeff:* (underdog) I'm a big boy now, not a child. If I'm old enough for military service, then I'm old enough to decide on my own morals. From now on, I'm going to live according to the situation, and not to rules, and if

I break some of your moral standards, it's because I find them useless and outdated. I'm not rocking the boat and bringing the family into disrepute—I'm just being myself for a change. That's just your catastrophic expectation—so shut up and leave me alone.

Old Jeff: Well, I don't know, it all sounds so dangerous.

New Jeff: Nonsense, I'm old enough to do my own thinking, and you'll just have to trust me. And you know you *can* trust me. I'm not likely to go off the deep end—I'm just not that kind of person. And in any case, remember, in the dream—there was one light that wouldn't go off however hard I tried. As long as that light is there, you needn't worry about me.

"Give me a child until it is seven years old," the Jesuits of old were reputed to say, "and I will give you the man." Jeff is a good example of how right they were—but hopefully his work with dreams will release him to some extent. With that particular topdog within, I doubt that Jeff will do much serious rocking of the boat—unless, of course, underdog is pushed to the end of his tether, in which case anything can happen. Morals were made for man, not man for morals, but it takes a hero to stand up for his needs. I hope Jeff will have the courage to put this particular topdog in his proper place—the servants' quarters—and live his life joyfully as his own master.

Whenever floods and tidal waves appear in dreams as topdogs I am not surprised to discover that the dreamer has been brought up in the Jewish-Christian tradition, knowing that this is God's method of punishing sinners. On the other hand, water is often a symbol of the unconscious and the emotions for many people, including myself. I am usually threatened by tidal waves in my dreams after controlling tears or anger during the day—a sure message to let them flow next time. The ocean of the unconscious becomes a tidal wave only when we refuse to let it flow, and our emotions threaten to overwhelm us only if we try to suppress them. The emotions are not a nuisance constantly distracting us from higher things, as many people in our society believe: they are the spring from which our vitality flows as we respond to the world, and we do not need to control them, for they function according to their own natural rhythm in the context of the situation.

PARALYSIS NIGHTMARES

In the periods of REM sleep, when most dreams occur, the body's motor system is immobilized, which is why we find it so difficult to cry out during a nightmare. (Sleep-walking and sleep-talking normally occur during non-REM periods, when the body is once again able to move.) There may be a connection between this paralysis of the body and the frequent experience of finding oneself unable to move in dreams, but it cannot be a straightforward connection, for if it were, we should be paralyzed in all our dreams, which is certainly not the case. By the same logic, it is a mistake to explain away paralysis dreams by saying that the bedclothes must have been too heavy. If the dreaming mind takes the trouble to pick up such physical sensations, it is always because the dream is expressing a feeling that something or someone is "paralyzing" the dreamer in his present life.

If your dream has no objective reference to some external person or event in your life at the time, then it is probably reflecting the fact that you are currently at impasse because topdog and underdog are pulling you in opposite directions with equal force. For example, when I was younger, I had recurring dreams of being chased by men and becoming paralyzed as they were about to catch me. At the time I had no idea what they meant, but it is now obvious that I was living in an impasse situation where my sexual needs were demanding expression, yet were stymied by a conscience that could not give permission. The dreams clearly showed underdog getting what he wanted and saving his face at the same time—for who can blame a victim of paralysis? And this was what I was doing effectively in my waking life, namely sending out unconscious sexual signals that would lead men to pursue me, and then holding them entirely responsible for the consequences. Had I been able to understand my dreams at the time, I could have spared myself a great many unpleasant encounters.

Herb, an advertising executive in one of our workshops, reported a dream in which he had taken an illicit afternoon off from work and was walking round the streets of New York when he came to Times Square and found it full of down-and-outs. A little man approached him and asked if he was on his way to work; when he replied that he was not, the little man said, "Oh then, you'll be wanting some of this—$20 for six," and he pushed some hash into Herb's hand. Herb, terrified that the police would get him for possession, threw down the package and tried to walk away, only to find himself totally paralyzed—and he woke in terror as the little man approached threateningly. When we worked on the dream it emerged that Herb's "Puritan-ethic topdog" had the catastrophic expectation that taking an afternoon off from work would inevitably lead to rotting in the gutter as a drug addict. Further work on the dream revealed underdog's determination not to be browbeaten whatever the cost: he was at the end of his tether in business life and was trying to find the courage to leave the rat race. As soon as Herb realized that his underdog was desperate and would sabotage all efforts at work, he determined to take the plunge. The last we heard was that he is taking a course in seamanship in preparation for sailing his own boat round the Pacific Islands—and there isn't a drug addict in sight!

Edgar Cayce had a theory that paralysis dreams indicated the need for a change of diet, and while I have not been able to find any evidence of this—the causes usually being psychological in origin—the notion is certainly worth considering. Most of us would benefit by a change of diet anyway.

RESOLVING NIGHTMARES IN WAKING FANTASY

The simplest way to deal with a nightmare is to continue it in waking fantasy, along the lines already illustrated in the foregoing examples. If there is a threatening character in the dream who is chasing, attacking, or imprisoning you, then give it a voice and ask it what it wants of you. If you have not been able to discover from the day's events whether it is a topdog or underdog, then dialogue in waking fantasy will soon reveal its identity. If it starts to scold or bully you, it is without doubt a topdog out to punish you for having given in to a need during the day, but if it whines and pleads with you to do something and threatens to spoil your plans if you don't, then it is an underdog in search of acceptance because you have in some way put it down during the day. Whether the ogre turns out to be a topdog or underdog, by giving it a voice and letting it speak for itself, you strip it of the fearful energy it had in the dream itself. The conflict is out in the open and you are able to deal with it.

There are, of course, some nightmares that do not fit into the topdog/underdog pattern, that are plain statements of a situation in your life. These, too, can be stripped of their terror by conducting dialogues in waking fantasy—and when this is done, we find that Chaucer was correct in that God does indeed turn every dream to good. Ann, the professor of psychology whose dream of being the first woman astronaut to the moon was described in Chapter 8, sent me a beautiful example of a horrific nightmare which, on further work, turned out to be a really loving and caring message from her inner self. She describes it as follows:

> I had a wonderful dream that illustrates how dream power can appear to be coming with frightful imagery when actually she comes with beauty and friendliness. I had just spent a lovely weekend with the current man in my life and so I asked dream power what I felt about him in my heart—that is, what do I feel about my own judgment in a love relationship at this time in my life?
>
> I dreamed I was vomiting large maggots and that my eye was being eaten by smaller maggots. My mother was standing by me during all of this, as I faced a mirror and vomited into the bathroom sink. My father was in a phone booth, wondering what on earth was wrong with me and phoning a doctor rather casually.
>
> Well, the dream was terrifying, repulsive—but when I acted it in a small group of friends and students, I soon learned that dream power's imagery remained friendly. I talked to the large maggots first, and the dialogue went like this:
>
> *Ann:* Who are you and what are you doing inside me?
>
> *Large maggots:* We are father maggots—derived from all the hurt and anger you felt about your father's rejection of you. It's a year since you worked on your motorcycle dream and confronted your father in fantasy. You got a lot of us out of your system then—but there are still some of us left inside. We've been here from childhood, so it's not surprising that it takes time for us to go. We came in your dream last night because you need reassurance that you are indeed getting rid of us, and all this coughing over your father is causing us to be thrown out of your system.

The motorcycle dream, which is described fully in Chapter 13, refers back to one of our dream groups Ann attended the previous year, in which she confronted her father in dialogue and came to understand something of his lifelong rejection of her. As she says, many large maggots were vomited up at the time, but this latest dream came to tell her that the matter is not completely closed, that she has not totally released her father and forgiven him, that some hurt and anger still remain, but also that she is slowly getting rid of it.

Ann then talked to the small maggots:

Ann: Who are you and why are you eating my eye?

Small maggots: We are young maggots feeding on your "I." We are young because we haven't been around long, only as long as your divorce. We'll go away from your "I" eventually, but we came to tell you that there are still some of us around, and you need to remember this and not fall in love too quickly. If you get yourself into a close relationship with another man at this time, we and the big maggots inside could spoil it, so be patient.

Ann: But what can I do to get rid of you? I don't like you eating my "I."

Small maggots: Oh, just keep doing what you are doing, sleeping with this current man, who is very good for you, but don't start thinking marriage yet. Just be natural and spontaneous about the whole thing and let it flow—this is healing to the "I."

Ann: But how can my "I" heal if you're feeding on it?

Small maggots: Oh, come on, Ann. You're not as weak as all that. Your "I" is as strong as a horse. It can contain us for a little while. Don't try and get rid of us by spraying or injections—that will only poison the system more. Just let it be for a while—your psyche is large enough to see this mourning period to its end. It will terminate in its own good time in its own natural way, and we'll be able to fly away on our own. Let it be—for the moment anyway.

The message to Ann was clear. The childhood hurt was a deep, inner wound which, at age thirty, is still festering, despite some very constructive dream work. The hurt over the divorce, on the other hand, is only two years old and has less seriously attacked the "I," the ego, or outer self-image. But even this needs time to heal. The dream warns Ann not to rush into another marriage just in order to save her self-image and prove that she is not a failure, but to take things easy and let the hurt clear up in its own good time. Ann concluded her description of the dream and its dialogue by commenting, "Dear dream power, thou art so careful and loving and protective of me."

RESOLVING NIGHTMARES BY REQUESTS TO DREAM POWER

In the case of Ann's maggot dream, although the imagery was horrible and frightening, she was able to continue it in dialogue and discover that its message was in fact kindly. Some nightmares, however, are so horrifying that they incapacitate the dreamer for further work. This happened in the case of Ann's "big, ugly ear dream," which she described in the same letter:

> I dreamed of looking in the mirror and seeing myself with a big, ugly ear with tentacles outreaching in all directions. Next to it on my cheek was a big ugly red mark, like a dessert mold. This scared me so much that I woke up in fright and couldn't go back to sleep for a long time, and the fearsome image persisted for several days so that I was quite unable to do any work on the dream. As I have had ear trouble before, I went to the doctor for a checkup as I feared going deaf, but every-

thing was fine. It cost me $30 for misunderstanding my dream! I then decided to ask dream power for a clarifying dream, asking it to send the same message in a less frightening way. So before falling asleep several nights later, I said, "Dream power, you scared me to death. Please tell me what that big, ugly ear dream is all about?

Clarifying dream: I am a college student and back in my childhood home. One of my current students, Betsy—a Jesus Freak—is coming to see me, so I put on my religious record —Jesus Christ Superstar—and hope this will bridge the gap between us, as I feel her present use of religion is neurotic and unhealthy. When she arrives, I call her Barbara, although I recognize her as my student, Betsy, in actuality.

Interpretation: The second dream gives me the triggering event of the first dream, which I had forgotten in the ensuing terror over the dream itself. At the time of the big-ear dream, I was grading the papers of my students and having an especially hard time knowing what to write in response to Betsy's paper on Jesus Freaks. I felt rejecting toward her because I think all that Jesus bit is terribly unhealthy, just as I now consider my own past orthodox attitude to religion unhealthy. (Before taking up psychology, I was in a seminary and later became a campus minister, but I'm so embarrassed about this that I never tell anyone about it.)

The big, ugly ear with outreaching tentacles is a beautiful picture of the way I am constantly on the alert for anything that smacks of neurotic religiosity, thereby hoping to avoid being branded with the religious stereotype myself, and to retain my image as the competent, rational, no-nonsense psychologist. The dream says clearly that, far from avoiding the "mold," my rejecting behavior is stamping it all the more firmly upon me, for it is common knowledge, even to the layman, that we project what we most dislike in ourselves onto scapegoats.

But my second, clarifying dream seems to suggest that I

should not be so hard on myself. My religious record—religious background—was a bridge leading from one religious stage to another. As a student I was curious and searching, and as a campus minister I was caring and concerned and tried to help others in their search for truth. While the outer form of my religion might have been a bit neurotic, its basis—the search for truth and goodness—was fundamentally good. As if to rub this message home, the dream shows me calling Betsy by the name of my beautiful friend in Boston, Barbara—telling me to look below the surface and recognize the basic beauty in Betsy and her present religious quest.

Result: This made me realize that I am still so scared of being stereotyped religious and goody-goody that I still go out of my way to be "unholier than thou"—smoking, drinking, and sleeping with boyfriends—probably more than I would do naturally. From now on, I'm going to try to accept my own goodness and spirituality, which does not mean being branded as a religious fanatic, and I hope this will enable me to accept the goodness lying beneath the surface in others. Only in this way will I become my own natural self—with no use for the big ear or the mold.

I also had a very fine conversation with Betsy. I told her about my dream, and in the course of our talk I realized that she really is a beautiful person. And she told me that she recognized that I had a kind of spirituality different from hers that she wished she could have. She said she felt my own spirituality was much more central to all my life than hers, more secular but more integrated into my life.

Reconciliation: I changed her grade on her Jesus paper from a "C" to a "B"!

The way Ann chose to work on these dreams did not lead to a specific topdog/underdog confrontation, though it is easy to see that one existed—a "psychologist topdog" constantly berating a "clergyman underdog" for being unhealthy, neurotic, and escapist and condemning him to lifelong exile from her personality on account of a few mistakes. Topdog was so terrified of the "clergyman" that he constructed the big, ugly ear as a kind of radar device constantly scanning the environment for hints of religious neurosis. The whole self-torture game had been sapping the energy Ann needed for her work as a psychotherapist, in which a sensitive, listening, accepting ("third") ear is essential. In this case topdog's lie was that Ann would immediately become a goody-goody killjoy preacher if the psychologist in her ever relinquished control—a notion so absurd that it is surprising she ever fell for it. Ann, like all of us, had indeed been neurotic if we define the word in Perls's terms as someone who cannot see the obvious.

RESOLVING NIGHTMARES BY MANIPULATING THE DREAM

A more sophisticated method of dealing with nightmares is to train ourselves to deal with hostile dream images when they actually threaten us in the dream itself. This is the method used by the Temiar tribe of Malaya, a division of the Senoi, who were first brought to the attention of the Western world by an Englishman, H. D. ("Pat") Noone, in 1931. They are a democratic, nonviolent, self-reliant people who live in harmony with nature and with themselves. The unit of Temiar society is the extended family and they live according to the law "Where a man has given his labor he has a share in the harvest," though each man receives not in proportion to his skill and labor, but according to his needs. Knowing that they have to contend with many hostile things in nature, they rely heavily on the help and goodwill of others. "Cooperate with your fellows," is a Temiar maxim. "If you must oppose their wishes, oppose them with goodwill."

Noone was so impressed by these peace-loving people, who

claimed they had not suffered a crime of violence or intercommunal conflict in several hundred years, that he wrote in 1932:

> Anthropologically speaking, such a tribal personality is unique and there is undoubtedly some scientific explanation for it. Personally I believe that the Temiar have hit upon some psychological trick in their educative process which enables the majority to reach this high degree of emotional adjustment as adults. I am not sure of my ground as yet, but I feel it has something to do with their dreams.

In 1934, he presented a paper entitled "The Dream Psychology of the Senoi Shaman" to the Oxford and Cambridge Faculty of Anthropology, in which he described the Senoi dream theory and techniques, later to be amplified by the American psychologist Kilton Stewart, who joined Noone in Malaya.

Nightmares, in Temiar dream theory, indicate the presence of hostile "spirits" within the dreamer's psyche which will become dissociated from his major personality and turn him against himself and his fellows if they are not promptly dealt with. As long as they are allowed to persist they tie up energy in wasteful psychic, organic, and muscular tensions. The children are taught from an early age to take their dreams extremely seriously as a major part of life, and much of their education is concerned with techniques for ridding the psyche of these hostile dream images. They are taught to carry some conscious awareness from the waking state into the dream state, so that when they are threatened by a hostile image, they will turn and confront it instead of running away. If the threatening character insists on fighting, then the child fights it to the death if necessary, calling on friendly spirits for help. The theory is that, once overcome in a dream, the hostile image becomes a friend and servant, and all the psychic energy tied up in the fight is released for creative living—and dreaming.

The parallel between this and the theory of nightmares I have been putting forward in this chapter is exact. If the hostile dream image is an underdog, we need to confront the impulses we fear in ourselves and discover that their terrifying qualities are no more than the fictions generated by the catastrophic expectations of topdog. On the other hand, if the nightmare is instigated by topdog, he will probably remain implacable even when confronted—or even try some new method of attack—and he must be subdued before his true status as friend and servant can be realized. I have managed to train myself over the past few years to follow out this program in nightmares on many occasions, and my experiences corroborate the Temiar view. If the hostile dream image is an underdog, confrontation almost always removes its hostility, whereas a topdog image usually continues to threaten and has to be defeated.

An example of making friends with an underdog in a nightmare occurred when I was being pursued one night (for the umpteenth time) by an enormous uncouth bruiser figure from the underground. I suddenly remembered my principle and turned around, not without some trepidation, and asked him what he wanted. He immediately doffed his cap and invited me into a sleazy café for a talk, whereupon he explained he had been following me for some time because he found me beautiful in comparison with his ugliness and wanted very much to marry me. I had the feeling he would turn nasty if I refused, and moreover I was beginning to feel sexually turned on by his brute physical strength but was worried about marriage because I thought we'd have nothing in common after the sexual attraction wore off. I woke up wondering what on earth to say to him.

This gave confirmation to what I had already suspected from working on earlier "bruiser" dreams, namely that he was an underdog representing my own brute strength and animal erotic energy, which my parental and spiritual topdogs had never really allowed me to accept in myself. (In Jungian terms I suppose he would be called a "shadow-contaminated animus figure.") I was struck with the parallel to the story of Beauty and the Beast, and keeping my fingers crossed that he would eventually become a handsome prince (not the insipid fairy-tale type), I had a chat with him in waking fantasy and accepted his proposal. I have never had these particular nightmares again and have felt much stronger ever since. I hope we shall live happily ever after.

Another case of a dream underdog turning friendly when confronted was sent to me by Marilyn, a correspondent in Philadelphia. She had been working with dreams for some time, trying to break free from childish acceptance of parental values, and dreamed of a burglar stealing all the furniture from her father's store under the pretense of cleaning the windows. Her first reaction was to call the police, but she found all the lines broken. The robber was now pursuing her because she had discovered his identity, and she fled to her childhood home, only to find it empty and unsafe as shelter because the locks on the doors were broken. She had no option but to turn and confront the robber—whereupon he somehow merged with her and they became one. On waking, she had no difficulty in recognizing the robber as her inner drive toward health and independence which was determined to make a clean sweep of her life by clearing out the old parental values (furniture), although one part of her still saw this as a criminal action. Once confronted, however, the robber ceased to be something fearful and destructive and was actually integrated with her other side in the dream itself, giving her a whole new freedom in waking life.

The willingness to confront and "embrace" the things you fear in yourself is the essential clue to integration, as both Jung and Perls realized, and this principle is at the heart of Temiar dream theory. Richard Noone writes in his book about his brother's life and death among the Temiar, *In Search of the Dream People:*

> To dream correctly, the child is told he must never be afraid. If he dreams of smoke, he must not avoid it as he would smoke during his waking hours because it stings the eyes. He must go boldly into dream smoke, for deep inside it he may find something of value, perhaps even the spirit of the smoke, which he can overcome and bend to his own will.

The parallel with Perls is particularly striking here. Fear may be rational in external life, but in the inner world it is fatal to run away from something frightening for victory lies in going directly into it. As Perls writes in *Gestalt Therapy Verbatim:*

> You never overcome *anything* by resisting it. You only can overcome anything by going deeper into it. If you are spiteful, be *more* spiteful. If you are performing, increase the performance. Whatever it is, if you go deeply enough into it, then it will disappear; it will be assimilated. Any resistance is no good. You have to go full into it—swing with it. Swing with your pain, your restlessness, whatever is there . . . use all that you fight and disown.

On the same principle, Temiar children are taught never to be afraid of falling in a dream, since it is merely an indication that the earth spirits are attracting you and have something to give you if you have the courage to face them. Once you have learned to let yourself go, a dream of falling often turns into a gorgeous flying dream, as I have been able to corroborate many times now in my own experience.

Our tendency to try to avoid the frightening encounter with ourselves in life is reflected in the fact that we usually break out of a nightmare at its most terrifying point, by waking up. Normally this happens involuntarily, but sometimes we "wake up" to the fact that we are dreaming in the dream itself, and jerk ourselves out of it with an enormous sense of relief that "it's only a dream after all." By doing this, however, we are depriving ourselves not only of a chance to deal with the nightmare creatively along the lines I have been describing, but also of a most remarkable dream experience—the so-called lucid dream in which you are no longer the victim of the dream process because you are aware of yourself as its dreamer. This remarkable state of consciousness is in my view one of the most exciting frontiers of human experience, and it can develop in all kinds of dreams. In fact, one of the most thrilling rewards of playing the dream game is that this type of consciousness, with its feeling of "other worldliness," begins to manifest itself much more frequently as self-awareness grows through dream work.

The West is only at the very beginning of exploring this fascinating area, which has been the main preoccupation of our own personal dream work for some time past, and I shall be glad to hear from anyone who has done serious experimentation along these lines. My present hunch is that the most constructive use of lucidity in a nightmare is not just to reprogram the dream story so as to give it a happy ending, as some people do—this may be an advance on just waking yourself up, but it is still an avoidance of the challenge in the nightmare story itself. The first step, after the realization that you are dreaming, is to confront the threatening dream image and see what happens, *in the conscious knowledge that you are a dream body and cannot be hurt.* If it is an underdog image, this could be the start of a fascinating adventure, as it might have been with my bruiser had I been lucid enough to accept his proposal on the spot. If it is a topdog image, on the other hand, you can launch yourself into battle with the knowledge that you can neither be hurt nor hurt anyone else—an equally important consideration in our society where topdogs so often resort to being hurt by our behavior when they can no longer control it by bullying. If the threatening image is too big or strong for a fight to seem sensible, then the knowledge that you cannot be hurt will still enable you to oppose it by passive resistance, as the following example of my own illustrates.

On the night before a TV interview on *Dream Power* in Newcastle in the north of England, I dreamed I was pursued by hostile Eskimos flourishing knives and hatchets. I suddenly became aware that I was dreaming and turned to face them. I falteringly asked them what they wanted of me, but they remained silent and threatening. As they were armed, I didn't relish a fight in spite of my lucidity, so I said in a trembling voice, trying to sound convincing, "You can't hurt me—I'm a dream body." As they lunged at me with their knives, I turned my head until they'd finished and then looked down at my side. When I saw all the knives and hatchets sticking in me, I laughed and said, "There, I told you—I'm a dream body and you can't kill me," at which they all looked very sheepish and walked away.

I was later able through dialogue to identify the Eskimos as my "north-of-England topdog" who berated me for writing a book on a frivolous topic like dreams when the people up north were still struggling to survive. I laughed at what I considered my own stupidity, until two days later when a reporter from the *Yorkshire Evening Post* held out my book and asked brusquely how I thought dreams could ameliorate the fate of the Yorkshire miners! Thanks to my dream, I was able to catch the feeling of hurt and recognize it for what it was—an assault of my inner topdog rather than an objective reaction to external criticism. And although it was triggered by a specific event in my life at the time, the dream has come to my rescue on many other occasions of criticism and attack, for when I remember those knives and hatchets in my side and the sheepish look on the faces of the Eskimos, I am often able to counter with humor instead of hurt confusion.

A similar case of a topdog who seemed too overwhelming to confront in open battle was given me by Nancy after attending one of our dream groups in which we had discussed lucid dreams. Her dream is a splendid example of a topdog trying more than one method to retain control. Having worked on a mother problem in the group and come to what she felt was some measure of resolution, she decided to ask dream power how she felt about her mother now. In a letter describing the result, Nancy wrote:

> In the dream, I was at the bottom of a hill with Rupert (my small son) when I suddenly heard shots. Then a bullet caught me full in the face—and I became lucid, perhaps because the bullet did not harm or kill me. Then a "thing" came around the corner at great speed—it looked like a mechanical dredger or digger with huge teeth—obviously bent on our destruction. Rupert was terrified, but I said (remembering your instructions), "It's only a dream thing, and it can't hurt us. Let it grind us if it wants." At this, it just disappeared, but a moment later Rupert and I were on the ground being given electroshock treatment by invisible agents. I told Rupert to lie still and reassured him that they could not hurt us. I felt a slight shock through my head, but my consciousness remained intact. Then the scene changed, and Rupert and I were flying through the air hand in hand having a delightful time. . . ."

Nancy added, "The dream certainly answered my question and painted an exact picture of how I see my mother. But the marvelous thing is that I *know* she can't hurt me any more—that the resolution I felt in the group was not just a head trip. The flying at the end of the dream and the joy on waking corroborate this feeling."

Last night, my dream tiger reappeared and chased me through a forest. As he leapt upon me and threw me to the ground, I suddenly became aware that I was dreaming. Instead of waking myself up from the nightmare, I realized that this would be a good opportunity to find out for sure what he wanted, so I pushed him off and said, "You're a dream tiger, and you can't hurt me, so let's be sensible and have a talk." For a moment he looked distinctly nonplussed, then obviously thought better of it and returned to the mauling. It then occurred to me that he was probably very hungry, so I "magicked" a large steak out of the air and told him to eat that instead. This he did with great relish, and I woke up as he sat down beside me licking his paws contentedly. In true forgiving undertiger fashion, he held no grudge against me for having starved him for so long, and I realized that regular feeding—allowing my animal nature and natural wildness a reasonable amount of expression—would make him my friend, guide, and protector for life. In fact, I have since been sufficiently lucid in several dreams to call upon my tiger for help when confronting an implacable dream enemy—which indicates a growing capacity to mobilize my own inner strength and resources in waking life instead of pushing the responsibility for my problems onto others.

Were I a member of the Temiar tribe, my tiger dreams would be taken as a sign from the spirit world that I was destined to become a shaman or dream authority, for in their tradition, according to Noone, everyone has some kind of guiding spirit that comes to him in dreams, and the tiger is the highest type of spirit or *gunig.* He pursues the dreamer until accepted, whereupon he becomes the spirit of protection and creativity for the whole tribe. In modern terminology, the tiger with its "fearful symmetry" is a vivid symbol of the power and wild grace in the unconscious, which seems threatening when we try to put it down, but if accepted will be our guide through the "forests of the night."

«12»

DEATH AND THE DREAMER

As long as you do not know how to die and come to life again, you are a sorry traveler on this dark earth.

—Goethe

There is an age-old belief that dreams can foreshadow impending death, but this is certainly not true of the great majority of dreams about people dying—if it were, the world would have been depopulated long since. Normally the dreaming mind uses death as a metaphor to express the fact that our feeling for someone, or someone's feeling for us, is dead, or that we have allowed something in our inner life to die—in just the same way as dreams use killing as a metaphor for resentment that wants someone to "drop dead," or for attempts to get rid of things we dislike in ourselves. On the same principle, if we dream of people who have already been separated from us by death, the meaning of the dream usually has nothing to do with whether *they* are still living beyond the grave, though they may be: they come into our dreams because they stand for something significant in our minds which urgently needs resolution. But the most interesting dream death is our own, for this indicates the death of some obsolete self-image, from which comes rebirth into a higher state of consciousness and authentic self-being. At this stage the metaphorical language of dreams seems to point beyond itself, to an experience which begins to overcome the fear of literal death by putting us in touch with a level of vitality that can transcend the demise of the body as we know it.

DEATH IN THE FAMILY

A dream of the death of someone close to us in our present life is very rarely predictive, though it usually has something to do with that person rather than with some aspect of ourselves. If it is predictive, it can almost always be traced to signs of sickness or imminent death which have been picked up by the dreamer's waking mind though he may not have been fully conscious of it. You may warn the person if it feels right to do so, but it is foolish to worry about such dreams or to cause anyone else to do so since the great majority of them are symbolic.

Popular psychology has made everyone familiar with the fact that dreaming of the death of someone close to you can indicate resentment against that person, a desire to "get him or her out of the way." Unfortunately, many people have taken this as an occasion to reproach themselves, when they have such dreams, for being nastier than they thought, which is the very reverse of the message the heart is trying to get through. Such dreams are usually warnings of neglected needs and feelings which will cause more trouble the more we try to suppress them. The best solution is to talk the dream over with the person concerned or to find some more creative and fulfilling life-style.

Not all dreams of this kind express resentment or aggression. Many people in our workshops, especially women, have dreamed of their partners dying and have traced the dreams to the feeling that the partner is moving away from them in some way—for example, by becoming more and more withdrawn and inturned. A dramatic case of this is given in Hannah Tillich's autobiography *From Time to Time,* where she tells how her husband, Paul, dreamed that their son had died, and then that she had died, when he himself lay dying. Evidently his heart saw his withdrawal from this world as his family departing from him.

The dream death of a partner can sometimes even be an occasion for rejoicing, evidence that on a psychological level a "happy release" has occurred. This was clearly the case with one of my friends who, at the end of a dream about the death of her husband, got into a car and drove away from the family home to explore the world—a sign that she had at last decided to allow her husband to be himself after twenty years' constant but totally ineffective struggle to turn him into the home-loving family man she thought he should be. By laying her image of him to rest, she had freed herself for a new life.

Parents very often dream of their children being lost or killed, and while these dreams should certainly be checked out for possible warnings to keep the child away from a dangerous swimming pool, high places, stairs, or whatever, depending on the details of the dream, the motivation in most cases is psychological. It can be resentment at the child's displacing the dreamer from the center of the family's attention (particularly common with fathers), in which case a family discussion is usually indicated. In the great majority of instances, however, the hidden villain who encompasses the child's dream death (or hurt) is a "parental topdog" who is punishing us for not coming up to some quite unrealistic standard of what a good parent should be. Dreams like this can be sparked off by the most trivial "lapses" during the day, like allowing the child to stay out a little later than usual or snapping at him because we ourselves feel hurt or angry. I had a personal illustration of how ridiculous the catastrophic expectations of this topdog can be when I dreamed of Fiona dying of syphilis. After a little thought I traced this to having rolled around on the floor with her the night before, which made my parental topdog come out protesting, "You shouldn't stimulate the child like that. She's getting all excited. She'll die of a sexual disease." Recognizing the old familiar voice, I told "my mother" to shut up!

DEATH OF A STRANGER (AND OTHERS WITH WHOM WE ARE NOT EMOTIONALLY INVOLVED)

Newspaper and magazine articles frequently describe uncanny cases of dreams that seem to be premonitions of the death of people at a distance, sometimes relatives or acquaintances of the dreamer, sometimes public figures like President Kennedy, but for every dream of this sort that seems to "come true," there are a hundred that do not. Elsie Sechrist, who firmly believes that dreams can sometimes contain precognitions, made the point at a recent A.R.E. lecture that her students so often call her to report dreams of her death that she would never put a foot out of the house if she took these as literal warnings. Instead, she recalls that when Edgar Cayce was alive his followers continually had dreams of his dying, but after he actually did die, they started having similar death dreams about his son, Hugh Lynn, who is still very much alive. She interprets all these dreams symbolically, on the assumption that spiritual teachers like the Cayces and herself stand for certain ideals which the dreamers are trying to follow but feel they have allowed to die in their own lives. I am sure this is the "correct interpretation" in the majority of cases, but I would add the cautionary note that the dream death could be a happy event rather than a sad one if the dreamer had been accepting the ideals in question from some kind of spiritual topdog inside himself, rather than from a true vocation.

A vivid case of a "premonition that wasn't" was given me by Craig, a TV executive on the *Tonight* show. Some time after the show had moved from New York to Hollywood he dreamed of a friend on the east coast driving over a cliff in her car. The dream was so vivid and frightening that he called her the following day, only to discover that she was fine. He interpreted the dream as a warning that he was in danger of losing all his old east coast friends if he did not keep in touch with them, and I noted this as a very striking case of a dream drama impelling the dreamer very directly to take the action urged by his heart. On the other hand, when Calvin Hall dreamed recently of an acquaintance plunging over a cliff in a car, he assumed the dream had no objective significance and tried to interpret it in terms of his own impulsive nature, until the news came through a day later that the boy had been very badly hurt in an auto accident. In his letter relating the incident Dr. Hall said that despite his skepticism about extrasensory perception, he had resolved in future always to get in touch with the person and warn him of possible danger after a dream of this kind. This seems to be a sensible precaution even though such cases are rare.

Sara, whose dream diary appears in the Appendix, called me long-distance one Thursday after a very vivid dream in which her friend, Martha, who was not intimately connected with her life at the time, had died on a certain date, the date being that of the following Friday. Sara wanted to know whether she should call Martha and warn her, or whether this might lead to anxiety on Martha's part. I said I thought there was no harm in calling Martha to see how she was and what her plans for Friday were, and to tell the dream if it felt right. In the meantime, I suggested that we try to interpret the dream in symbolic terms. Asked to describe Martha, Sara said she was a mixture of earthiness and spirituality, qualities she much admired. In fact, Sara had chosen her last job, in which she worked at housing low-income families, because she felt it gave expression to her own spiritual and worker sides, but she said that conflicts within the organization were damping her enthusiasm and killing all her feeling for the work. When the following Friday came and went without the real Martha dying or in any way coming to harm, it seemed evident that the dream had used her as a symbol to warn Sara to do something about her own "deadening" situation.

Several months later Sara dreamed of receiving a news clipping about the death of a friend she had not seen for many years, whose name was Tommy Arnold. During the day she had met with the leader of a Transactional Analysis group she was thinking of joining and had not liked him at all. She interpreted the dream in terms of initials—T. A.—and decided not to join the group as her feelings for it were now dead.

While she was on a visit to us, Sara dreamed that Pope John XXIII was responsible for the death under the Nazis of the thirteen-year-old Jewish girl Anne Frank, because he had not come to office in time to save her. Although Sara knew quite well in her waking mind that this made no sense in terms of real politics, it somehow seemed perfectly logical in the dream. She could not at first relate the dream to anything in her day, so we asked her what happened in her own life at age thirteen. She replied that her family had moved to the city, where her father had a big public job which kept him away from home practically all the time. The memory of her loneliness and longing for him brought tears and revived the anguish she had felt at being deprived of his love and attention.

Associating to Anne Frank, Sara described her as a beautiful, warm, happy, trusting, innocent child—all the things Sara herself had been before her own symbolic "death" at age thirteen. She said that in order to gain her father's love and approval she had repressed or "killed" the soft, loving, dependent side of herself and become cool, efficient, independent, and unemotional, just like him. She then realized that the dream was telling her that she felt her life could have been saved if only this cold "Father" had been replaced by a warm, loving "Papa" like John XXIII, whom Sara much admired. Since on the previous evening we had been discussing Sara's current feelings for her father (who, of course, has never become the "Holy Father" she would like him to be), it seemed clear that our conversation reminded her heart of this longstanding resentment. We believed the dream had chosen these two particular characters because when Sara arrived she found me (Ann F.) at home alone while my husband (John) was out giving a theological lecture, taking her back to memories of times when her father had been out working in the evenings instead of being at home with her.

DREAMS OF THE DEAD

I have received hundreds of letters and questions after lectures, asking whether a dream about a dead person is a real visitation of that person's surviving spirit or just an ordinary dream. I have no personal experience of this since I have lost very few people who have been close to me in life. In my dreams, and in those of our dream group members, dead people usually appear in a symbolic drama, as Anne Frank did in Sara's dream, to make some special point or to represent some quality meaningful to the dreamer. On the other hand, Elsie Sechrist, who has more knowledge of such dreams than I have, and also more personal experience of her own, does believe that some dreams of this kind really are visitations by the dead.

The commonsense question to ask, it seems to me, is why

the surviving spirit of a dead person should want to come back in someone's dreams. If the dreamer has been worrying about whether a loved one has survived death or not, the visitation could be intended as reassurance, but it is not very sensible to think of this happening more than once or twice. Beyond that, a visitation would presumably be for the purpose of conveying some kind of message to the dreamer about his own life, in which case the need is to find out what the message is and decide what to do about it—and from this point of view, it seems to me to make no practical difference whether you treat the dream as a visitation or a message from your own deeper self. For example, Don, who attended one of our Florida workshops, told us how he dreamed a great deal about his grandfather immediately after the old man's death and in subsequent years has continued to dream of him from time to time whenever he feels in need of the kind of wisdom his grandfather brought to him while still alive. Don is a firm believer in life after death, but he does not feel the need to connect these dreams with the actual spirit of his grandfather. He thinks rather that he invokes the old man's image as a way of contacting the resources of wisdom inside himself, his own best personification of a guru within or "wise old man archetype" (as Jung called it).

I would certainly agree with Elsie Sechrist when she writes, "Do not fear conversation with the dead in dreams." She suggests that a departed loved one who appears distressed in a dream may be calling the dreamer to pray for his soul, and this could indeed be a valid dream message for any dreamer whose religious beliefs include the idea that the dead need our prayers to help them on their soul's journey. Elsie's experience and mine coincide in finding that the commonest message of the departed is to ask us to stop grieving. In a lecture given at the A.R.E. she quoted the case of a woman who dreamed of her dead mother drowning in a pool, saying, "Daughter, these are the tears you have shed for me. Let me go for I am drowning in them."

This is an eminently healthy message whether you take the dream as a visitation or a communication from your own heart, but it is easier said than done, as I know from the many letters I have received. I usually advise the correspondents to continue the conversation with the dead person in waking fantasy, possibly with the help of another person or small group, to discover why they cannot release him, even when they have been assured in dreams that he is still well and happy. I will illustrate the kind of problem that often emerges by quoting a typical dream and dialogue from one of our groups.

Ruth, age fifty-seven, who had lost her husband, age sixty-five, about six months prior to this dream, told us that immediately after his death she had dreamed of him constantly, usually in past situations of happiness. Naturally she spent the following day crying because it had all been just a dream. Then one night she dreamed that he told her he would have to go away, that he had his own things to do, at which she burst into tears and pleaded with him not to, saying that she could not cope without him. We asked her to finish the conversation in waking fantasy, and the dialogue went something like this:

Ruth: Please, Al, don't go off and leave me again. I need you. There's so much we still have to do—all those things we never did in life, so much unfinished. I want to tell you I love you and try to make up for all the times I hurt you and disappointed you and . . .

Al: But you've done that, honey. I understand, I really understand.

Ruth: No, you don't. If only you'd given me some warning —I'd have been nicer to you. But you didn't tell me (sobbing) and now it's too late, it's too late . . . how was

I to know? Now I'll feel guilty for the rest of my life (angrily) and it's all your fault. You should have warned me.

Al: But, honey, I didn't know either. I didn't want to die.

Ruth: And what's more, you went off without putting all your affairs in order. Oh, I know you thought you had, but you hadn't. And I'm a lot poorer than you thought —so much in taxes. And the income tax, the papers, the move . . . suddenly having to cope with so much. You've no idea how hard it's been.

Al: I'm sorry about that, Ruth. I should have taught you how to cope with things.

Ruth: But I don't want to have to cope with things. That's your responsibility. Why do you think I married you? I don't want to cope, I don't like it.

Al: But, honey, that's life—we all have to grow up sometime. I just wish I could have prepared you for it.

Ruth: (crying) But I need you, Al. I'm lonely.

Al: You'll make a new life, new friends. I know you will. You *can* if you want to.

Ruth: But that's just the point. I *don't* want to. I want *you.* You are my life . . . so much invested in you. I can't let you go.

[I interjected here to say, "Try saying, 'I *won't* let you go.' "]

Ruth: I won't let you go . . . I *won't* let you go (angrily) you're goddamn right, I won't let you go. Why should I? I was always there when you needed *me.* It's your turn to be here when I need you (shouting) but, no, you're in an exciting new place with new experiences ahead of you, in all those many mansions the minister talked about at the funeral. And afterward he has the gall to spend two hours telling me how happy you're going to be and not to grieve for you, you with your nice new body. What about me? That's what I wanted to ask him. What about me, in this old body with arthritis? No new life in store for me—only loneliness, sickness, and old age. Who's going to grieve for me? Not you, not anyone. Well, bully for you, that's all I can say, bully for you.

At this point in the dialogue, Ruth was pounding her knees with her fists, and someone threw her a cushion. She got down on her knees, pounding it furiously, embarking on a tirade of all the hurt and anger she had bottled up inside for twenty years—all of which boiled down to the fact (or feeling) that she had given him her life and now he was walking out on her, leaving her nothing. He had broken a contract that existed only in her heart. As she wept and pounded, she shouted, "Give me back my life, Al, give me back my life."

Sensing this to be a very healthy statement, I asked Ruth to return to the chair and reclaim her life from Al.

Ruth: (softly) Give me back my life, Al, all those stolen years. Give me back my youth, my laughter, my joy.

Al: They're yours, honey. They were always yours. I didn't take them. You gave them to me. You seemed to want to give them.

Ruth: (crying) I did, I did. I loved you, but I need to reclaim myself. I have to reclaim my independence, my strength, and my love—because I might need it for someone else (surprised). Well, I might (defiantly). I can live without you. Hey, I *can* live without you (laughter from group).

I asked if she could tell him good-bye now.

Ruth: (softly) Good-bye, Al. I'll miss you (crying) so much we could have done. Well, it can't be helped. I'll still think of you sometimes—often—but I'll let you go. Good-bye—have a good trip (smiling). Take care.

Ruth's predicament is one that repeats itself time and again in our groups. So often the bereaved feel guilty about all the things they could have done in life and didn't, and resent the fact that the deceased has put them in this position. Then behind the resentment lies the anger and rage at being deserted, left to cope alone in a lonely, hostile world. The head knows quite well that these feelings are absurd, that the dead person did not die on purpose, but this kind of logic does not touch the heart, and unless we can bring these irrational emotions out and really work through them in some way they chain us to the past and to our memories, locking up the energies we need for making our new lives. Sometimes they even make us physically ill and incapacitated.

Ruth's reaction could in no way be called pathological, for her husband had died only six months previously and it was natural that she should still be dreaming and grieving about him. On the other hand, if she had not released her emotions in the dream group, she might well have been dreaming and grieving about him years after his death, which *is* pathological. Dr. Vamik Volkan, a psychiatrist at the University of Virginia Medical Center, has noted that many of his patients who have frozen into permanent grief have certain common dreams in which the dead person is alive but caught in a life-or-death struggle. It seems to me that this dream struggle depicts the dreamer's love-hate relationship with the dead person and consequent ambivalence about the death. Volkan has developed what he calls a "regriefing therapy" for such pathological mourners, in which they are helped to relive the original grief process and express all the ambivalent feelings connected with it, this time ensuring that they accept the death emotionally as well as intellectually.

In our culture we are doubly inhibited from facing and expressing our feelings of anger and resentment against the dead person, because in addition to being irrational these feelings are also selfish, and hence at variance with our ethical standards. It is bad enough to feel selfish at any time, but to feel selfish about someone who has died seems so awful that we usually dare not contemplate it. And, as Ruth's dialogue brought out, the Christian rites connected with death do not help us here, in their bland assumption that our concern is primarily with the fate of the departed and our sorrow entirely due to altruistic love of the other person. We could use the psychological sophistication of some of the more primitive religions which had a place not only for loud expressions of grief, but also for getting out "unfinished business" of anger and resentment against the dead person. This does not imply a "lower" view of human nature than the Christian one—merely a realistic recognition that love of self and love of others are inseparable and that attention must be given to the heart's hurts and anxieties on its own behalf before love, grief, or concern for others can emerge.

Our moral or religious topdogs probably have the catastrophic expectation that to give way to selfish feelings would cause them to get such a hold on us that we should never feel anything else, but experience shows this to be quite false. Arthur Janov brings this out vividly in his book *The Primal Revolution,* when he describes how his father's death took him back to the "primal" experience of childhood anger and despair at never being accepted for himself. Janov's particular brand of psychotherapy consists of bringing this experience into conscious awareness, so although he was a grown man when his father died, he put himself to bed for four days to go into "primal feelings."

> What I cried about was the father I never had. I was crying for me. That's all any neurotic can do when he is filled with Pain. Each day I felt an entire strip of feeling that was apart from any other feeling. One day it was, "Say I'm good before you die." Another day it was, "Be my good daddy." Still another Primal was, "Don't go now. I still need you," and so on. Each feeling consumed hours of memories and new feelings and then insights.
>
> When I finished with the four days, which left me so weak

I couldn't get out of bed, I could cry *for him* for the first time—for the tragedy that was his life. Until my need was out of the way, no feeling of mine could be objective and external. After that cry for him, I have never cried since, nor have I mourned. It feels like it's over. Not that I don't remember it all with sadness.

Had I not had Primals to fall back on, I would have had a general depression and mourning that might have gone on for weeks and months. It never would have been resolved because I would have had no idea why I was really crying and mourning. I still needed a good daddy. . . . It may seem heartless and cruel to say that my first act after his death was to cry for me, but I had no choice. I didn't will it that way. It happened that way because my needs took precedence over any other thing, as they always do.

Later, I wondered how it felt for my father to know he was fatally ill, surrounded by all those doctors, being alone and frightened. I started to console myself that it wasn't so bad. After all, he was old. But I knew this was rationalization. He was a scared little boy, bewildered by what life had come to for him, and he died in his childhood.

The impasse in which we refuse to experience feeling is a kind of living death, and that is what we condemn ourselves to when we cling to the departed. There are other dynamics that can lead us to this continued dependency besides those revealed in Ruth's case, although this kind of "unfinished business" is by far the commonest reason. Sometimes, however, a "faithful-forever topdog" is at work, telling us that we must be heartless, unfeeling morons if we do not allow our lives to be permanently overshadowed by loss of the loved one. If this kind of sentiment is expressed by visions of the dead in dreams, I am willing to assert categorically that they are not visitations from the next world but projections from the dreamer's own mind and must be told firmly to get out of his life.

Possibly the most difficult to overcome of all the reasons for refusing to let the dead go is the fear that grief and loss will be overwhelming, that without the other person there will be just a great hole of nothingness in our lives which we cannot endure. Perls quotes an example of this in *Gestalt Therapy Verbatim:*

> There was once a girl, a woman, who had lost her child not too long ago, and she couldn't quite get in touch with the world. And we worked a bit, and we found she was holding onto the coffin. She realized she did not want to let go of this coffin. Now you understand, as long as she is not willing to face this hole, this emptiness, this nothingness, she couldn't come back to life, to the others. So much love is bound up here, in this coffin, that she rather invests her life in this fantasy of having some kind of a child, even if it's a dead child. When she can face her nothingness and experience her grief, she can come back to life and get in touch with the world.

Facing what we believe to be nothingness, the void—that is the essence of every impasse. Clinging to someone who has died is just one more manifestation of the basic neurotic phenomenon of the inability to believe that we can have any meaningful existence at all except in terms of some self-image with which we have become identified, whether it be of wife, mother, upstanding citizen, dutiful son, or whatever. To experience the death of the self-image and to discover that beyond it lies not nothingness, but new life—this is at the core of all religious teaching and all psychotherapy. But, as Perls writes, "We'd rather maintain the status quo: rather keep in the status quo of a mediocre marriage, mediocre mentality, mediocre aliveness, than to go through that impasse." And Jung said, "People will do anything, no matter how absurd, in order to avoid facing their own souls. They will practise Indian yoga and all its exercises, observe a strict regimen of diet, learn theosophy by heart, or mechanically repeat mystic texts from the literature of the whole world—all because they cannot get on with themselves and have not the slightest faith that anything useful could ever come out of their own souls."

Dreams of one's own death almost always reflect the fact that we have reached the point of being willing to relinquish our old roles and self-images, and for this reason these are often the most important dreams of all.

THE DEATH AND REBIRTH OF THE SELF

The children of the Temiar tribe are taught never to fear death in a dream, from whatever source it comes, but to go through it to the joyful discovery that they survive and emerge with new strength. To quote the description given by Kilton Stewart, "When you think you are dying in a dream, you are only receiving the powers of the other world, your own spiritual power which has been turned against you, and which now wishes to become one with you if you will accept it." In Gestalt terms such a dream shows the dreamer giving up the struggle to refuse everything that does not conform to the self-image imposed on him by a topdog and recovering the hitherto rejected energies of underdog. A beautiful illustration was provided by Gail, who had worked with us for some time in an ongoing group, and she tells the story in her own words as follows:

> In the first dream of the night, I am taking a French exam. and I cheat. Every time the French teacher goes out of the room, I take a book from the shelf and look up words. She never discovers my cheating, but suddenly the time is up, and I've only half completed my paper.
>
> On waking, I lay in bed and wondered what on earth the dream meant, and my immediate association to French was sex. So I pondered how I had cheated on sex. We hadn't made

love the previous evening—in fact, I'd actually put David off when he suddenly got horny in the middle of our yoga. I told him I felt sexy too, but I thought we should finish our yoga. Actually, I didn't want sex at all, but I just couldn't admit it. I mean, after all, a *real* woman wants sex all the time . . . that's supposed to be her pleasure and her life . . . if she doesn't, then she's frigid. (I'd just been reading Xaviera Hollander's books, and she always seems to be horny!) Then it all clicked into place. I'd cheated by pretending to be the totally sexy woman—and as far as I knew, I'd got away with it, by insisting that we continue our yoga—putting higher things before my own desire! And then I asked myself, why do I have to go to these lengths? Who wrote on the wall that a real woman is always sexy? Well, there are the Pussycats and Fascinating Womanhood, and all those journals that tell you to do it even if you don't want to . . . wow! I fell back to sleep wondering whether I could give up my self-image of the sexy Gail or not.

In the next dream I am cornered in a game of knives, chased by men with knives, and find myself in a large room. I run out of the door but find it leads into the corridor I've just left. The men approach, and I am cornered. There's no escape. I turn to face them, and as they plunge their knives in, I accept my death.

Then the scene changed, and I find myself at home. I hear a child's cry, and I suddenly realize in horror that I have a small child locked up in my garage and I must let her out. Heavens, she must have been there for years and I forgot all about her. I rush to the garage and open the door with apprehension because she must be very angry with me. A small, dark girl about nine hurls herself at me and hugs me fiercely. I hug her back, and just as I'm wondering why she's not angry and resentful, she says, "I'm a big girl now," and keeps repeating this. She looks like an American Indian, and she talks a lot—but I let her because she's been locked up for so long. Then suddenly we are in a restaurant, and she is talking away about her travels in such a grown-up, intelligent way that the people

> at the tables around all smile and listen to her, as much as to say, "What a lovely child!" I look at her and think, "Well, for such a deprived little thing, she's remarkably happy and perky and entertaining," and I wake up really loving her.

When Gail first recounted these dreams in the group, we felt there was no need for further dialogue: this was one of those cases where the dreams had done it all. Here was the age-old theme of death and resurrection—in Gail's case, the death of a phoney sex image and rebirth into authentic self-being as the lovely child. As Gail explained, the child was not beautiful in any ordinary sense at all—in fact, she was rather dark and plain—but her enthusiasm, sparkle, and wit drew the attention of all those with whom she came into contact. Gail had locked this dark child—her underdog or shadow—in the garage at an early age when the world told her that the only way to succeed was to be sexy and subservient to men. This meant that she had also locked up her innate intelligence, independence, and joy. The dream itself completed the integration, and we felt it had a message for ninety percent of women in our society who are brought up to consider themselves as sex objects (French women make no secret of it!) with no worth other than this. For all the women of the world who have built up sex images for themselves it seems like death to let the image go—but it is really death into life more abundant, as Gail soon realized from her dream.

Every underdog or shadow figure is a child who has been locked up in the garage or cellar because we have been brainwashed by topdog's catastrophic expectations of what will happen if we allow his needs to be met: in Gail's case, she really believed that if she did not play the sex game, she would become the ugly, frigid, uninteresting person she so disliked. This was the monster she believed she was locking up, little realizing that the "natural" child was far more attractive and warm—and more truly sexy—than the sex object she was trying so hard to be. In a similar way I had locked up my own warm, dependent feelings in the cellar many long years ago, believing I had imprisoned a horrible, clinging old witch constantly asking to be loved—until a dream showed me opening the door to a lovely lost little kitten. Sara had done the same thing with her own soft feelings so disliked and feared by her father, and could not bring herself to accept this side of her personality until she started dreaming of a loving, playful, mischievous puppy urging her to romp with it. Even the nastiest things in ourselves have positive qualities, and they manifest as fearsome monsters of the id only so long as we keep them under lock and key, but when we have the courage to open the door, out steps a child, a kitten, a puppy, all eager and waiting to become part of our life. I am not propounding a theory here: I am speaking from personal experience as a kind of pioneer who has explored the forest and found it friendly. The life-force is never vindictive or destructive if you befriend it.

The flow of energy, joy, and creative power that often comes with this experience seems to me to require the language of religion to give it adequate description, and I was fascinated to find my feeling corroborated from an entirely different viewpoint by Thomas Harris in his book *I'm OK—You're OK.* Using the terminology of Transactional Analysis, he suggests that the essence of true religious or mystical experience is the overthrow of the inner Parent so that the Natural Child can emerge from behind the Conditioned Child and cooperate with the intelligence of the Adult in a life no longer alienated from the ground of being. Harris writes:

> It is my opinion that religious experience may be a unique combination of Child (a feeling of intimacy) and Adult (a reflection on ultimacy) with the total exclusion of the Parent.

> I believe the total exclusion of the Parent is what happens in *kenosis,* or self-emptying. This self-emptying is a common characteristic of all mystical experiences. . . . I believe that what is emptied is the Parent. How can one experience joy, or ecstasy, in the presence of those recordings in the Parent which produced the NOT OK originally? How can I *feel* acceptance in the presence of the earliest *felt* rejection? . . . I believe the Adult's function in the religious experience is to block out the Parent in order that the Natural Child may reawaken to its own worth and beauty as a part of God's creation.

One very important corollary of this in my own experience is that the process of releasing and reowning the inner child seems to be accompanied by a change in the person's attitude to literal, physical death, as reported by mystics great and small down the ages and in all cultures. In my own life, and in the lives of those with whom I have worked, I have not encountered a sudden, dramatic, once-for-all change, but rather the dawn and gradual growth of a new attitude, which is nonetheless highly significant for living.

A very important factor here is the emergence, in the dream life itself, of new kinds of consciousness which begin to suggest *in experience* that we may be able to grow into modes of being not dependent on physical life as we ordinarily know it. One such experience is the "lucid" dream, in which we become aware in the dream itself that we are dreaming and are in a world which, by some strange paradox, is both ours to manipulate yet somehow not dependent on our manipulation. Sometimes this experience is accompanied by a vivid flash of awareness that lifts us literally into another "dimension" of greater clarity and joy, where the dream body seems literally to leave the sleeping physical body and go on all manner of strange travels, flying, passing through solid objects, and altogether encountering the strange new universe with high ecstasy. This experience, known in religious and occult literature as "astral projection" or "out-of-the-body experience," is closely related to the commoner erotic flying dream and seems to correspond to a special kind of "high" in waking consciousness, the high that comes when the energies of the natural child are allowed to flow.

These very special dreams, which now form about ten percent of my own personal dream record, are to be the subject of my next book—but I want to make it clear here that the new attitude to death I have described does not *depend* on these: to think of it in that way is probably to put the cart before the horse. The key factor seems to be something much subtler, yet down to earth. I discovered this for myself when I decided not long ago to try Edgar Cayce's experiment of asking my dreams for enlightenment on my own feeling about death. In the dream that followed I found myself on my hands and knees on the floor in my childhood home, where two young men whom I thought of as Mafia types were about to kill me brutally for having accidently disobeyed some order they had given me. Waking in horror, I immediately associated the Mafia with "the family" and realized with astonishment that my whole view of death was colored by my childhood experience of punishment within the family, with no less than death the penalty for disobedience. We all have our own unconscious views of death, and these fall like a shadow between the human spirit and its natural existential sense of eternal being. In other words, the overthrowing of topdog, the dethroning of the inner Parent, is indeed, as Harris suggests, the key factor; once that is done, the sting of death is removed and the sense of growing beyond the limitations of the life of the physical body is slowly but surely added to us.

This is why I find myself in emphatic agreement with Harris in distinguishing genuine religious experience from something that at first sight often looks like it but is actually its polar opposite—the attempt of underdog to find peace by giving in to topdog, usually through some form of religious conversion in which he expresses regret for having rebelled so long against parent-type restrictions and resolves in future to conform to what "God" wants. Harris writes:

> There is one kind of religious experience which may be qualitatively different from the Parent-excluding experience we have just described. This is the feeling of great relief which comes from a total adaptation to the Parent. "I will give up my wicked ways and be exactly what you (Parent) want me to be." . . . Salvation is not experienced as an independent encounter with a gracious God but as gaining the approval of the pious ones who make the rules. . . . Freud believes religious ecstasy is of this sort: the Child feels omnipotence by selling out to the omnipotent Parent. The position is I AM OK AS LONG AS. The conciliation produces such a glorious feeling that there is a hunger for it to happen again. This results in "backsliding," which paves the way for another "conversion" experience. . . .

I would add that the same holds good if we try to solve our inner conflicts in ordinary life by giving in to topdog in a nonreligious context; underdog cannot and will not conform to an alien pattern, however sensible and worthy it may seem, for he is the essence of what *we* have to become, the image of God *in us,* distorted into something ugly only by our topdoggy rejection of him.

To accept the dominion of topdog over our inner world is literally a fate worse than death, as I had brought home to me vividly in another dream which occurred some time ago. In it someone had proved by physics that if certain lines of force came together in a certain pattern, the person caught in them would be condemned to live forever alone in an empty universe. As so often happens in dreams, no sooner does the thought occur than it actually takes place—and I found myself immobile, imprisoned in a web of constricting forces from which there was no escape. It was an appalling experience with nothing at all but these strange lines of force and the sun glaring down merciless and cold. For the first time in my life, I understood what Paul Tillich meant by the "vacuum" and what Fritz Perls meant by the "death layer"—for there could have been no better picture of the living death that comes from trying to freeze oneself into the inertia of a self-image. For the first time that I can recall, I longed for death. And then, after an eternity of suffering (or so it seemed) in the midst of the nothingness, I saw in a glass before my eyes the reflection of the sun, which slowly started to move, to expand and contract. I felt warmth on my body and knew that my ordeal was over. This too was a picture of what has happened in my life as a result of dream work. As Perls puts it, to fear annihilation is to remain stuck in the void, but when we accept what we fear and go into it, the desert starts to bloom. The empty void becomes alive: the sterile void becomes the fertile void, and we are reborn.

«13»

THE TRANSFORMING SYMBOL

Problem-solving is like patching holes in a rotten boat; for each patch applied, two more leaks spring up. There are times when a way out is needed that is not available to logical patching techniques. There are times when we need a way beyond rotten hulks, a way not for restructuring a new boat or even a serviceable life-jacket, but rather some sub-mariner's way through a sea of confusion to new terrain.

—Joseph Chilton Pearce

A common theme of many ancient myths and legends, from all parts of the world, is that when the hero has overcome the giant or dragon who has been holding the land in thrall, he finds some kind of token—a precious stone, a magic feather, or similar talisman—which forever after will help him summon up strength when he is exhausted, have vision when all is dark, and find the way out when he is imprisoned. Anyone who makes a serious effort through dream work to confront the inner tyrants of the soul (topdogs) will make a similar discovery if he stays on the lookout for it. For, on such occasions, the dreaming mind often throws up images which seem to have a mysterious, almost magical power to evoke the deepest energies and creative resources of the personality, not only at the time of the dream, but for months or even years afterward whenever topdog tries to make a comeback. Sometimes the "transforming symbol," as I call it, comes directly out of the dream, while on other occasions it emerges in the course of working on the dream in waking fantasy, but either way, it provides the dreamer with something like a mental charm that can help him in most extraordinary ways in future times of trouble. Yet the symbol need in no way be "extraordinary" in itself, as the following examples will show.

EASY RIDER

A remarkable example came from a very short dream brought to one of our workshops by Ann, the professor of psychology whose later dreams have already been described in Chapter 8 ("first woman astronaut to the moon") and in Chapter 11 ("vomiting maggots" and "big ugly ear"). At the workshop where we first met her, we had as usual suggested on the Friday evening that group members put a request to their dreams, and Ann had done so, asking for enlightenment about her most pressing, nagging problem—her relationship with her father.

She had left her husband some time ago and at the time of this dream was still wondering whether or not to divorce him. The process of making the break had become agonizingly drawn out, and she knew this was largely her own fault in that she was reluctant to let him go, but the insight had not helped her do anything to change the situation. She suspected that her inability to make a clean break with her husband had something to do with her feeling of being a failure as a woman —a feeling originally derived from her rejecting father. In fact, she had been undecided about whether or not to go home this particular weekend instead of attending the dream group, in order to have a confrontation in depth with her father.

She tells the story of her dream and subsequent events in her own words:

> *Request:* Dream Power, should I go and see my father? It's taking me so long to get a divorce. Is it the old father-hurt making it take so long? Could I reconcile something if I went home to see my father?
>
> *Dream:* I am getting on a motorcycle. My father is standing there saying, "You'll never make it. You'll never be able to ride a motorcycle." I get on and ride beautifully, hair blowing in the wind, and super control around curves and with high speeds. I enjoy it. I ride it and drive it well. I get off the motorcycle and my father is standing there saying, "See, I told you that you'd never be able to ride a motorcycle."
>
> *Interpretation:* Father will never be able to tell me he is proud of me. I have to find that voice in me that says what he would say if he were able, namely that he secretly does respect me, and does not respect my passive and dominated mother. He does love me but cannot say or show it. When I get in touch with that voice underneath the lying topdog, I cry because I need so much to hear that secret truth.
>
> *Gestalt work:* The most helpful work was my dialogue between "Tomboy Ann" and "Feminine Anna." That dichotomy within me was never before reconciled. The aggressive part of me protects me from being crushed like my mother in her passivity. The feminine or softer part of me is the gentleness that men appreciate. But I need both these parts of me to be a whole person. My father never allowed me to be either—he despised my feminine side and called me fat and ugly, and he belittled my masculine side, which drives me to achieve. The result was impasse—all my energy locked up in the stupid self-torture game and very little left over for celebrating life.
>
> *Transforming symbol:* All the old hurt and anger is an extremely powerful energy source within me when transformed into a motorcycle—a symbol of energy, power, and creativity.

The motorcycle has been a tremendously helpful transforming symbol in my decision to take my Ph.D.—especially now that I have the understanding and insight that I can ride my motorcycle to creativity in jeans or a skirt—in both my masculine and feminine aspects.

After-realization: The real father continues to be a super son-of-a-bitch. I went home on a Thanksgiving visit and he was totally punishing, name-calling, mean, and rejecting. It is *impossible* for him to show any of the respect I'm sure he feels for me. *Remedy: Divorce Father!* Try to remember what he would say if he could, and above all to remember that I *can* ride a motorcycle, whatever he says. I must protect myself from the father within, topdog who calls me worthless and ugly, and work out all my unmet needs and superanger by learning to *love myself.* And if I have any more nonsense from topdog, I shall get on my motorcycle and run him down—and run him down—and run him down—until he accepts and loves me, until I accept and love myself.

It is hard to do justice in words to this remarkable experience. When we first began to encounter these almost magical symbols in dreams, we were reminded of Jung's famous concept of archetypal symbols that emerge from dreams in times of crisis to point the way to psychological growth—except that in our experience the transforming symbols are hardly ever "archetypal" in the sense of being related to the age-old mystic images of occult and religious traditions.

In Jungian writings it seems as if the important dreams marking progress on the path of self-realization always involve sticks or swords emblazoned with ancient script, great trees standing in circular temple courtyards, mythic beasts rising from the ocean depths, doves descending on altars and suchlike, which has led many readers to conclude that unless their dreams are like this they must be relatively unimportant. Nothing could be farther from the truth. Ancient symbols usually occur in dreams when and only when the dreamer is already interested in ancient symbolism, and their meaning can often turn out to be quite mundane. One of our college students of Jungian background dreamed of being swallowed by a huge fish and interpreted the dream as a possible encounter with a Christ-symbol from the deep unconscious, although this meant nothing to him in terms of his current life. Further work revealed that he felt decidedly overwhelmed and "devoured" by his oversize girlfriend, whom he described as a typical Pisces! So it is very important to bear in mind that most life-changing dream symbols are quite ordinary objects derived from everyday life, like Ann's motorcycle, and could be missed if the dreamer has been led to think that major dreams must contain traditional archetypal images.

Ann's motorcycle would lose rather than gain in significance by being treated as a modern version of an ancient symbol like a chariot, or Pegasus, the flying horse. To begin with, there was no real equivalent to the throbbing energy of the internal-combustion engine in earlier times, yet this is a very vivid symbol of a certain aspect of psychological energy, combining in a single image the swiftness of the horse, the roar of the bull, and the beat of the pulse, adding the further quality of the motor vehicle itself, a product of human ingenuity and creativity that goes faster and packs more power than anything mankind knew in pretechnological days. But it would be quite wrong to think that the symbol of the motorcycle carries a universal meaning, for it will always have its unique and personal meaning to the dreamer—in Ann's case, expressing her feeling of being in perfect control of her own "masculine drive."

One of my own transforming symbols is a big, flashy American car, which first appeared in my dreams at a time when I really needed to get in touch with my energy, "drive," and "bigness," having been prevented from doing so in the past by a species of British topdog who insisted that I crawl around being mousy, humble, and self-effacing (obviously with the catastrophic expectation that I would become like one of those *awful* Americans!). I love my big, flashy American car and use it, among other ways—like Ann—to run down topdog whenever he starts yapping at me with his unthinking clichés.

On the other hand, Al, one of our students, also has a car as a transforming symbol, but his is a convertible and its message is quite different. Before this symbol appeared in his dreams, Al had been terrified of his own aggressive urges, believing that any expression of them must inevitably result in violence, so he pushed them away, pretending they didn't exist. The result was, of course, that they manifested in rather nasty destructive behavior in devious ways, and Al lost a great deal of natural productive energy. The convertible came to tell him, "Look, Al, it's not necessary to defend against these urges all the time. You can put the hood up when necessary, and take it down at other times. Try to live according to the situation, not a rule. I will protect you from danger when the situation requires it, but it's also fun to feel free and ride with the wind, in touch with nature and the elements."

A CLOTHING FOR THE SOUL DIVINE

Al's transforming symbol corresponds directly to what Jung called the "transcendent function"—in his view, a central feature of the psychotherapeutic process—namely, the emergence of something from the unconscious that provides the individual with a higher point of view from which an apparently irreconcilable emotional conflict can be transcended. (In Perls's terminology, this would be a force that enables us to get through the impasse). In his introduction to *The Secret of the Golden Flower,* Jung writes:

> The greatest and most important problems of life are all in a certain sense insoluble. They must be so because they express the necessary polarity inherent in every self-regulating system. They can never be solved, but only outgrown. . . . This "outgrowing" . . . on further experience was seen to consist in a new level of consciousness. Some higher or wider interest arose on the person's horizon, and through this widening of his view the insoluble problem lost its urgency. It was not solved logically in its own terms, but faded out when confronted with a new and stronger life-tendency. It was not repressed and made unconscious, but merely appeared in a different light, and so, did indeed become different. What on a lower level, had led to the wildest conflicts and to panicky outbursts of emotion, viewed from the higher level of the personality, now seemed like a storm in the valley seen from a high mountain-top. This does not mean the thunderstorm is robbed of its reality, but instead of being in it, one is now above it.

Jung stated that the transcendent function could sometimes emerge as an image from a dream but might also be worked out in waking fantasy on the basis of material thrown up in a dream (or under hypnosis or in a moment of artistic inspiration). I first discovered this possibility for myself when I obtained a "conflict-transcending" or "creative sidestep" transforming symbol in the course of working on one of my dreams.

The dream took me back to the home I shared with my first husband, where I was trying to mend an old, dark gray blanket. The more I tried to sew it together, the more it fell apart in my hands. I tried to clear up the mess but the pieces kept sticking to the floor. Then Ellis Amburn, the publisher, walked in, and I said, "Oh, everyone has been telling me to get in touch with you," at which he smiled and seemed pleased.

I needed no dream to remind me that several people had

told me to get in touch with Ellis, who had made the original contract for *Dream Power* in the United States but had since moved to a senior position with another publishing firm and, from what I'd heard, was making a success of it. I also had no practical problems of mending on my hands, but we had been talking about my former husband the previous evening, and I had been expressing regret that my attempts to remain friends with him after the divorce had failed. This suggested a fairly obvious meaning for the blanket, as a symbol of the relationship I was still hoping to patch up into something whole after its disintegration, only to find that it deteriorated more the harder I tried. This interpretation was corroborated by my recalling that he and I had initially got together after he held my hand under a blanket at a Theosophical camping weekend.

The dream seemed to present a clear objective message to stop crying over spilt milk and "get in touch" with Ellis—a double symbol in my mind of the successful side of myself. However, this excellent advice was no revelation—I'd been telling myself to get in touch with my own new successful life for a long time, as indeed the dream made clear by my saying, "*Everybody* has been telling me to get in touch with you." So I looked for a deeper message in the dream, perhaps a message about why I had never achieved this eminently sensible objective, and the key feature here seemed to be the way the pieces of blanket kept sticking to the floor and refusing to be cleared away—a sure sign of an underdog refusing to be wiped out.

I conducted a dialogue with the blanket which went something like this:

Ann: Look, you can see I'm trying to mend you. Why don't you cooperate instead of making a mess all over the floor?

Blanket: Can't you see I'm beyond repair? I'm falling to pieces.

Ann: Then I'll just have to throw you out.

Blanket: You can't do that to me. I'll stick to the floor. I'll hide bits of myself in the furniture. You can't get rid of me like that.

Ann: Well, you give me no option. If you won't be mended then you'll have to go. It has to be one or the other.

Blanket: You just try! You won't succeed. You can't get rid of me.

It seemed that I was at impasse—with topdog determined to clean the mess up one way or another, and underdog digging his heels in and refusing to move. I recognized my own voice as that of my mother, who all my life has nagged me for untidiness both in the house and in the mind, berating me for not being totally together and perfect (presumably as she believed herself to be). In "her" mind, there seemed to be nothing in between, as any hint of accepting less than total tidiness conjured up catastrophic visions of total chaos. I had obviously picked up this notion and incorporated it into my own life, and in this particular situation had been wasting an enormous amount of energy in striving for a "perfect, no-mess" relationship one way or the other with my former husband. With this insight, I continued the dialogue to see what, if anything, emerged:

Ann: Look, I don't want to go on with a stupid self-torture game forever. I do have a new life to lead and I need all my energies. I know enough psychology to understand that if you refuse to move, there's nothing I can do about it. So I'll try asking you what you want.

Blanket: Well, I can't be the original blanket ever again—I'm too far gone for that—but there are still parts of me that are good.

Ann: But you're so dark and gray and unattractive.

Blanket: I may be dark and gray, but I *am* warm. If you put me together with that lovely bright material left over from that dress you made Fiona, we'd have a super patchwork quilt, with my dark pieces highlighting the colors.

This suggestion came straight off the top of my head, while I was trying to get as closely as possible into the feel of the blanket image without interpreting it, but I could see at once that it made exquisite psychological sense. Fiona, the child of my first marriage, is the epitome in my life of light and spontaneous joy—but it's just not possible for life to be as perfect as that. The perfectionism which tries to sweep away all experience that cannot be made directly positive and good is the enemy of growth, not an aid to it. Blake made this perfectionism the archvillain of his great prophetic poems, personifying it as a character called Urizen who freezes growth everywhere because he seeks "a joy without pain," whereas the truth is (in Blake's own words) that:

> Joy and Woe are woven fine,
> A Clothing for the Soul Divine.

The dream enabled me to leave my futile efforts to "patch things up" with my first husband, while at the same time accepting the life I had with him, and even the pain of our separation, as integral parts of my total life experience. Far more important, however, is the fact that the image of the patchwork quilt, which came partly from a dream but had to be synthesized in the dream work, remains with me as a powerful transforming symbol, enabling me time and again to break free of perfectionism and embrace the "negative," painful, dark aspects of experience as integral parts of the garment of life.

TREES OF LIFE

Sometimes a transforming symbol emerges gradually over a series of dreams. A good example was provided by Christine, a housewife and former scholar with two grown-up children in college. On her husband's death a few years previously, she had taken up the study and practice of Eastern religions, and had even visited India and Tibet in the hope of finding her guru. She returned to England disappointed but continued to practice yoga and meditation, and had become quite a leader and teacher of various Eastern disciplines at the time of her dream. She had not worked much with dreams because the teachings she followed dismissed them as mere wandering thoughts to be superseded by the pure light of consciousness—but she was sufficiently intrigued by a dream fragment to ask us what it could mean. In it she was driving her car and found the rear view blocked by a large bushy tree on the back seat. Her only association was a certain tree she had been fond of in childhood, but she could not link this with anything in her life at the time of the dream. We suggested that she talk with the tree, and the dialogue went something like this:

Chris: What are you doing in the back seat of my car? You're blocking my view and we might have an accident.

Tree: For heaven's sake, you must have put me there—and I must say I don't like it at all. A car is no place for a tree. Look, my roots are drying and withering for lack of earth and water. I can't live unless my roots are in the earth.

Chris: Well, what do you expect me to do? I'm far too busy to bother with you at the moment.

Tree: If you don't plant me soon, I shall die.

Chris: That's too bad. You're not my responsibility. I have my work to do. Look, why do you bother me now just when I'm really getting into higher states of consciousness?

Tree: Because I need to be planted *now.*

Chris: I hate to see a lovely tree die. I like trees—but I've got my own thing to do, and besides I'm not a gardener. I wouldn't know how to set about planting a tree. But I'll make a deal with you. If you'll just get out of my way and let me get to my destination, I'll plant you then.

Tree: That's never, and you know it. You get one place and then you're onto the next, driving yourself all over the place—driving to reach this state and that state, and do this and do that. Well, I'm going to block this window and stick in the back seat here until you slow down or have an accident. You must plant me—this is an emergency.

At this point it dawned on Christine that the tree was her own spiritual growth, which was dying for want of a firm basis in the earth and of free-flowing emotion (symbolized by water). This astonished her, for she had always believed that natural physical life and its desires have to be put down if one is to achieve enlightenment, just as the tree in her dream seemed a threat to progress. When she found that the disciplines she had applied to rise above her earthly attachments failed, she had driven herself even harder to reach the goal, little realizing that this only served to "block her vision" even more. Underdog was at the end of his tether, and the message of the dream was clear; unless Christine paid attention to her physical and emotional needs, some kind of "breakdown" was inevitable.

In an extension of the dialogue, in which the tree confronted one of Christine's former teachers who talked of the need of the spirit to free itself from earthly fetters, the tree retorted:

Tree: Do you think you've been put on the earth for nothing? Do you think you have nothing to learn from it? I am your true spiritual growth—not just nature—the tree of life. With my roots deep in the earth, I learn its secrets and can convey them to the heavens; and with my branches high in the air, I learn the secrets of the sky and convey them to the earth. I bring the secrets of the worlds together—body and soul—and I provide a home in which nature's creatures can grow, as well as producing life-giving fresh air for them.

Christine was clearly coming round, in her mind as well as in her heart, to a new and more positive appreciation of natural life, and I was reminded of Jung's statement that "Great innovations never come from above; they come invariably from below, just as trees never grow from the sky downwards, but upwards from the earth," and Teilhard de Chardin's notion of consciousness evolving by enriching itself from "the sap of the earth."

The dream and dialogue must have crystallized the message for Christine, for a few months later she called us triumphantly to report another tree dream. This had taken place after a day of planting trees in the garden of a new home, an event which had obviously provided her heart with an occasion to express some very positive thoughts. In the dream a very special tree had arrived from the past and she had dug a deep hole in which to plant it. When it was firmly in the earth, she had shared with the tree (in the dream) a feeling of gratitude and elation that stayed with her long after she awoke. "I actually felt what it was like to be a tree," she said excitedly, "and it was beautiful!"

That was not the end of the story, though, for topdogs are not so easily eradicated. Some months later, Christine dreamed of a branch of a tree covered with caterpillars and awoke with a sense of horror as she dashed to get the insecticidal spray. The dream had obviously been sparked by seeing real caterpillars on an oleander bush in the garden, but she knew the dream's meaning was symbolic. She had spent the whole day doing nothing but a little gardening and reading and had enjoyed the sunshine and the birds. In fact, she had just sat for several hours feeling at peace with the world—a new experience for her. It was only when she went to bed that a vague feeling of guilt overtook her and made her wonder if she hadn't been slacking spiritually. (This reminded me of my friendship long ago with a Bahai pioneer on the island of Sicily. When I took her to the mountains one day to show her the natural beauty and grandeur of the area, and asked if she were enjoying it, she replied, "Yes, very much—it's quite magnificent—but I feel I'm wasting time: I should be at home praying!")

Christine's heart clearly thought she was letting the spiritual side down by indulging her lower needs, for it depicted her spiritual growth (the tree) being devoured by caterpillars. When we asked her to talk to the caterpillars, they started by whining in typical underdog fashion that they too had to live and meant no harm. She retorted in a very topdoggy way that this was too bad, but her tree was far more important and she would have to kill them. Then suddenly, on inspiration, she gave the tree itself a voice, and said, "Wait a minute? What's the panic? I'm a big, strong tree. A few bugs aren't going to kill me, you know. And don't forget that if I feed them, they will eventually become butterflies."

This turned out to be one of the most important insights in Christine's life, for although she had overcome her topdog sufficiently to recognize that spiritual growth is derived from, not opposed to, the earthy life of nature, she was still occasionally caught out in the old topdog habit of considering laziness, sensuousness, and other "lapses" as bugs preying on her growth. Possibly the most important lesson psychology has to teach us is that such lapses—temper, depression, overeating, laziness, and so on—are always protests from the deepest center of our being against our efforts to force ourselves to grow in some one-sided fashion. The lapses may be "bad" in themselves, in the sense that they often seem to be destructive, but the remedy is never (except in short-term emergencies) to try to overcome them by control. We need rather to listen to what they have to tell us for our greater health and vitality, remembering the wise dictum that all evil is potential vitality in need of transformation, and that the "bugs" of our psyche are our potential butterflies. Laziness is often the protest of a soul that needs more relaxation, and it can become the beautiful virtue of serenity; anger can become passionate caring and concern; fear can become necessary caution; and greed is a sign of failure to give full expression to the desire for growth itself. Depression shows enormous energy locked in by conflict.

The tree could easily be called an archetypal symbol by anyone who wanted to do so, since the world's myths abound in mystical tree symbols—the Tree of Life that appears at both ends of the Christian Bible, Buddha's tree of enlightenment, the Great World Ash Yggdrasil of Norse legends, and many others. Christine was not unaware of this, for she had studied mythology, but she insists that the important thing about *her* tree is that it has its "roots" in her own personal life. She tells me that the symbol now helps her not only to cope creatively with mental "bugs" like depression, moodiness, or sloth, but is effective also in dealing with the bugs of physical illness, which in many cases come upon us simply as the body's protest against being neglected or driven too hard by topdogs.

"THE TRUTH SHALL MAKE YOU FREE"

A transforming symbol with a meaning very similar to Christine's tree came to Don, who attended one of our ongoing dream classes. In a very short dream fragment he saw his beautiful, expensive stainless steel knife covered with rust, which in the dream itself he knew to be impossible. Don reiterated this when we asked him to have a conversation with the knife: "My eyes must be deceiving me," he said, "because you're made of stainless steel and *can't* rust," but when he took the character of the knife, he retorted shortly, "There's nothing wrong with your eyes, and I *am* rusty." As Don could not resonate to the notion that he was "getting rusty," and as he had already accepted the fact that he was not always as "stainless" as he would like to be, we left the dream at this point in the hope of further clarification during the course of the week. At the next class he reported another dream in which his knife had appeared to him even more rusty and crumbly than before.

Don was inclined to relate the dream to his lifelong involvement with Christian Science, whose teachings he interpreted in terms of the conviction that man, being made in the image and likeness of God, *cannot* manifest either physical or mental disease or disharmony. In other words, man is like a stainless steel knife that *cannot* rust. While Don's conviction had worked in his own life, he had difficulty in communicating this conviction to others, and his dream seemed to be telling him that his lack of success might spring from an internal failure to convey his conviction from his head to his own heart. But Don steadfastly resisted the notion that forty years of a life lived creatively and productively in accordance with these principles was a mere head trip, and pointed out that if it were, his unbelieving underdog would have sabotaged him long ago. He also pointed out quite correctly that the dream depicted a *perfect* object as imperfect, corroborating his heart's deepest conviction that man was indeed a perfect being. Don felt that the dream had some other message for him—something to do with the rust that was manifesting as part of his heart's reality and which refused to be put down as an error of perception.

Some weeks later, after the group had stopped meeting, Don had a further dream which sent him back to the knife dreams with a renewed conviction of their importance. It occurred in a semi-waking condition in the morning and was a vision of a street with big oak trees behind which stood large houses. A voice, speaking very clearly, said, "Donald, what I am trying to tell you is . . . ," and then after a pause for emphasis, "in my house are many mansions." Don was very impressed with this, because no one ever normally calls him Donald. He felt it was telling him that the clue to the clarification of his faith lay in some way of expressing the fact that the all-in-allness of God allows of many different ways of experiencing the perfection of the life process, including the experiences we call corruption, evil, and illness. Don also linked this with the work we had done in the dream class on topdogs who create havoc by trying to make us live by phoney self-images.

Because we were working intensively on the self-image problem at this time, we held a special session with Don to work further on his knife dreams, and at this point it became clear that they did indeed show the way to the missing factor of the equation in Don's thinking. As Don put it afterward, the clue lies in presenting the Christian Science conviction of perfection, not as an *alternative* to the ordinary view of evil and disease (as if one were trying with gritted teeth to convince oneself that "pain isn't real") but in using it as a complement to the ordinary view, like the complementary images in a stereoscopic picture which come together in a single focus to provide a vision in depth of the situation.

In psychological terms illness of mind or body has to be seen in a "positive" way, as a protest by the life process against a vision of ourselves which is too limited, a warning that we are denying some essential aspect of our being by trying to impose some false (topdog) image of what perfection should be. By refusing to let the life-force flow naturally (refusing to allow a rose to be a rose) we are quite literally choosing to experience life as a sickness rather than as health. When we insist, like Christine, on treating things that "bug" us as alien forces to be sprayed out of existence instead of recognizing them as signals from the deeper body/mind self about some unnatural strain we are imposing on ourselves, we actually *make* the "bugs" dangerous by forcing them into stronger and stronger protests. The willingness to tune into one's whole self, rust and all, is the true science of health, and "cure" comes, not from denying dis-ease, but by expanding consciousness to the point where we understand the reason for it—and then we can choose to let it go or not.

I have no idea whether this kind of formulation would be acceptable to official Christian Science teachers, but there seems little point trying to "cure" a person if the dis-ease is unconsciously "doing something for him" or if he is in some way "enjoying bad health." This was obviously the case with a close friend I knew long ago who contracted tuberculosis and spent several years in a very pleasant sanatarium in the country, away from his helpless and demanding widowed mother and dreary work. Under the sanitarium conditions (beyond his control) he was able to enjoy the care and attention lavished on him by nurses and friends, free of guilt and recrimination. Having little or no psychological knowledge at the time, he and I both persisted along Christian Science lines with the help of a practitioner, for consciously he wanted to get well—but without success. It was not merely that he needed a change of life-style (the practioner was well aware of this) before he could contemplate a return to the world; he needed to change his rigid and limiting self-image of the dutiful, upright, loving son who had taken on the sole responsibility for a widowed mother and fatherless family. It was this self-image that he could not release, and he remained sick. As Sidney Jourard writes in his book *The Transparent Self:*

> It is difficult and anxiety-provoking to change what one is doing, to change or reinvent one's way of interacting with others; powerful forces from within and without tend to restrain change, and so most of us keep up the way of life that has been slowly "doing us in." Therefore, we become sick, and it is usually with some measure of surprise. It is still an unsolved question why the sickness is "physical" for some and "psychiatric" for others. I am coming to suspect that those who are often physically ill are people who commit "altruistic suicide" by slow degrees. They are slowly destroying their bodies, as it were, for the preservation of their roles and the social systems in which they regularly participate. They are victims of their sense of duty. The psychiatrically ill seem to resemble rebels without courage or effectiveness.
>
> Being sick is a temporary respite from the dispiriting conditions of our existence up to the onset of the illness. Incidentally, if it seems to a patient that his usual life, the one that made him sick, cannot be changed, he may never get well. Why should he? Or if he does recover but then resumes his usual life, he'll be sick again before long. As he leaves the hospital, we could safely say, and mean it, "Hurry back."

For Don, this insight into the deeper causes of dis-ease, and into the ultimate nature of Reality, came as enlightenment of mind-blowing dimensions, as depicted by a "verification" dream the night after our conversation. The dream showed him at the airport seeing off his sister-in-law, as he had been doing in waking life before coming to us for discussion. In the dream a small airplane was flying overhead when it was tripped by a high wire. It bounced over and fell on some lower wires, whereupon it burst into an all-encompassing white light that filled the whole world. On waking, Don felt that dream "power" (high wire) had "tripped him up" in his usual line of thinking (flight of ideas) and caused him to explode into new understanding no longer too far "above people's heads." He wrote later that "This has been a wonderful insight into my most inner being and a culmination of at least forty years of study."

«14»

THROUGH THE DREAM GATES: ESP AND ALTERED STATES OF CONSCIOUSNESS IN DREAMS

If the doors of perception were cleansed every thing would appear to man as it is, infinite.

For man has closed himself up, till he sees all things thro' narrow chinks of his cavern.

—William Blake

The notion that dreams foretell the future, or are vehicles for visions and direct mind-to-mind communications, is age-old and still so widely held that many people expect a book on dreams to be entirely devoted to this theme. The dreaming mind's capacity to ignore the ordinary laws of space, time, and logic invites the supposition that it may have powers altogether beyond those of waking life, and in recent decades credence has been given to this notion by a number of psychologists who approach the subject from a scientific rather than an occult point of view. Many psychotherapists, for example, have documented uncanny instances of their minds apparently being read or their future actions foreseen in their patients' dreams, and since the discovery of REM periods in

1952, attempts have been made to study dream ESP under controlled conditions in the laboratory, as described by Ullman, Krippner and Vaughan in their book *Dream Telepathy.**

The main conclusion to emerge from all serious studies of the subject is that whatever paranormal powers the dreaming mind may possess, they are quite unimportant from the practical point of view in comparison with the "normal" function of our dreams, which is to give insights for daily life. Dreams certainly *can* give us important practical warnings but these can almost always be traced directly to the heart's subconscious detective work based on small impressions or subtle vibes, picked up by normal means during the day, which the waking mind was too busy to register or perhaps did not want to know. Dreams that seem to give warnings of a more dramatic and paranormal nature "come true" only in a tiny minority of cases; for every uncanny story that gets written up in the press, there are thousands of equally vivid dreams about friends dying, air crashes, floods, wars breaking out, or presidents being assassinated which are never literally fulfilled. And it is common knowledge among parapsychological researchers that the great majority of reports which really do seem to show ESP involve incidents that are totally trivial in themselves—like dreaming of a flamingo in the kitchen and finding one there on waking.

So while the question of whether or not ESP of any kind occurs in dreams is undoubtedly interesting scientifically, the popular belief that dreams are important *mainly* from this point of view is quite misplaced. The most important thing about dreams is that they do not have to "come true" to *be*

*Researchers commonly distinguish three kinds of ESP, or extrasensory perception: telepathy (direct mind-to-mind communication), clairvoyance (the mental vision of events at a distance), and precognition (prevision of the future).

true; they are all true anyway as reflections of the dreamer's life and problems. And it may be that we should bear this in mind when thinking of new ways of conducting research in this area—perhaps by looking for possible connections between apparent ESP and the heart's concerns at the time of the dream. This chapter is concerned mainly with my own studies undertaken from this point of view, which have led me to formulate a hypothesis of ESP in dreams based on the topdog/underdog model of the personality.

Parapsychological research on dreams has reached an impasse, since we can never know whether ESP takes place in the dream itself or entered the mind in waking life too quickly to be consciously registered; like any other stimulus picked up during the course of the day, it could remain dormant until the dreaming brain replays the day's events in depth. J. B. Rhine, the great pioneer of ESP research, constantly warned against difficulties of this kind which arise whenever one is trying to study a phenomenon without having any scientific idea of its mechanism. The fascinating experiments carried out at the Maimonides Medical Center in New York by Ullman and Krippner fall prey to this difficulty, for even when an agent appears to succeed in transmitting an image to a sleeping subject during REM periods, we can never know whether this is true on-the-spot telepathy or precognition on the part of the dreamer, who foresaw the whole experiment during the course of the day without consciously realizing it. This criticism is all the more serious because Ullman and Krippner claim to have demonstrated precognition in certain subjects. When an impasse is reached in scientific research—as in any life problem—it is necessary to sidestep the issue and ask new questions, and I hope that my contribution will be useful in indicating possible new directions in ESP dream research. In fact, this is an area in which every serious player of the dream game can participate, by keeping a regular dream diary, noting cases of apparent ESP, and observing possible patterns which emerge from them.

ESP AND THE DREAM DIARY: THE LEGACY OF J. W. DUNNE

One of the most famous of all systematic studies of ESP in dreams was carried out by an engineer, J. W. Dunne, who describes his experiments in *An Experiment with Time,* published in 1927. He was interested in only one kind of ESP, namely precognition. He began by noticing a number of incidents where his own dreams seemed to pick up events from the future with sufficient detail and regularity to make pure chance unlikely, although the events themselves seemed to be quite abitrary and had no significance for his life at all. This led him to hypothesize that the dreaming mind stands outside space and time, picking up events from the future and weaving them into a dream story in much the same way as it uses events from the past. He accordingly decided to keep a regular dream diary to check out this hypothesis and persuaded a number of his friends and acquaintances to follow his example.

His results convinced him that the dreaming mind uses past and future incidents in a precisely symmetrical way, with images from tomorrow occurring with the same frequency as images from yesterday, images from next week occurring with the same frequency as images from last week, and so on. This led him to elaborate his famous theory of "serial" or multidimensional time, which inspired several popular plays by J. B. Priestley. I do not know of any work that has corroborated Dunne's ideas, possibly because I have never known anyone with the patience and persistence to carry out his experiment systematically. As Dunne himself remarked, it is difficult enough to spot the minute details of incidents and thoughts that have gone into a dream from the *previous* day, but "to notice that a resemblance between a waking event and a past dream is worth following is like trying to read a book while looking out for words which might mean something spelled backwards. The mind cannot keep that up for long." I should be very interested to hear from readers who have persevered with Dunne's experiment with time, as my own attempts collapsed under the strain.

Today his book stands as an important monument in the history of research in dreams mainly because it established that a systematic dream diary is essential for any significant research in this area and pointed to the sensible conclusion that ESP (if it occurs at all) is likely to turn up in dreams in much the same way as ordinary waking perceptions. A dream diary is particularly important in showing the frequency with which we dream about certain things: if I dream of a flamingo in the house every other week, the coincidence of finding one there some day is not very impressive. As the Roman orator Cicero said two thousand years ago, at a time when educated and uneducated people alike took the occurrence of "supernatural" events for granted:

> From the visions of drunkards and madmen, one might, doubtless, deduce innumerable consequences by conjecture which might seem to presage future events. For what person who aims at a mark all day will not hit it? We sleep every night, and there are few on which we do not dream; can we wonder then that what we dream sometimes comes to pass?

If your dream diary shows that you have recalled one or two dreams a night over several years and have never on any other occasion dreamed of a flamingo in the house, and your record of the previous day's events and thoughts showed nothing about flamingos, then you have a really striking coincidence that looks like ESP. In itself, it is not enough to prove the existence of ESP, but it is a start—and research on any less secure basis is useless, even in the laboratory.

For any dreamer doing personal research in this field outside the laboratory, a dream diary is essential both for collecting as much material as possible and as a safeguard against the mind's well-known tendency to "write up" its memories after the event to fit an interesting possibility. Many people have told me that they dreamed of an unknown man or woman and then actually saw them the next day—which may certainly seem uncanny to the dreamer but carries no weight with a serious researcher unless there is a detailed description of the stranger in the dream diary. Experiments have shown that witnesses will often pick out someone as a criminal from an identification lineup when the real criminal is not there at all, and even when none of the suspects closely resembles him.

Another way in which this tendency of the mind expresses itself is the self-fulfilling prophecy. A woman told me that whenever she dreams of cats she finds money the next day, "not just sometimes but *every* time." Since she did not keep a dream diary, there was no way of telling how often she dreamed of cats, but I was able to discover by careful questioning that on one recent occasion when she had such a dream, she immediately started hunting for a $20 bill she had lost in the house. After another cat dream, a neighbor had unexpectedly asked to buy one of her chairs and she had agreed—but on any other occasion she might well have refused. I also suspect that her memory exaggerated the number of these coincidences, a fact which her dream diary would have revealed.

Because of the mind's proneness to self-fulfilling prophecies, I think we should be very cautious indeed about claiming apparent ESP on the basis of the interpretation of a dream as distinct from its literal images. If I dream of an airplane crashing in flames which looks like the Douglas DC 9 I shall be traveling in the next day, and I learn after completing the flight safely that some of my shares have "crashed" on the stock exchange, it is pushing the river to claim this as a likely case of ESP, since I could have done just the same had I received news of my uncle Douglas dying or seen Mike Douglas "blow up" on his TV show. If I dream of a bucking horse, and the next day receive some money (bucks), this is no more significant than receiving a letter from my friend Buck, or from my cousin who lives in Bucks County. (This means, of course, that many cases of real ESP may be missed when an image presents itself symbolically rather than literally—which all adds to the difficulty and confusion inherent in this field of research.)

My own studies so far have been carried out mainly with my own dreams, which I have been recording systematically for ten years, and with John's, which he has recorded for about half that time, also with the dreams of a few friends who keep regular dream diaries. I have concerned myself only with those incidents of apparent telepathy, clairvoyance, or precognition in dreams that make a spontaneous impact and have not followed Dunne's example in *looking* for possible corroborative events. Nor have I included symbolic dreams which have several possible meanings, even though I have been sure of my own interpretation. My aim is not to collect statistics to prove that ESP occurs, but to consider whether, in cases where it seems likely, it has any discernible psychological meaning.

UNDERDOG'S PSYCHIC RADAR

Freud, in spite of a strong prejudice against anything smacking of the occult, became convinced of the reality of ESP from frequent uncanny incidents in his work with patients. While some of these concerned dreams, and others indicated a capacity for mind reading in waking life, Freud believed he could detect a common pattern in all of them, namely a strong anxiety or need on the part of the patient which he was unable to communicate openly. For example, one day when Freud was anticipating with great pleasure a reunion with his old friend Dr. David Forsyth from England, his patient suddenly declared that his own nickname was Mr. Foresight. Freud interpreted this incident as a show of infantile jealousy on the part of the patient, meaning, "You see, I too am a Forsyth: you should be just as glad to see *me* as him."

The repeated occurrence of incidents like this led Freud to urge psychoanalysts to pay special attention to any cases of apparent ESP on the part of their patients because they seem to reveal important unconscious conflicts which might not otherwise come to light. They might also be useful to the analyst in calling attention to his own unresolved problems. Dr. Jan Ehrenwald, in his book *New Dimensions of Deep Analysis,* describes many such cases, a typical one being that of a dream brought to him by a patient called Ronald.

Ronald described in some detail a house beside a stream with a flight of stairs up one side and a sunny little green field on the other. This field, he understood, belonged to the Warner brothers, and he was forbidden to use it. A wall of grayish black rocks surrounded the field and there was a hill in the background. Ronald interpreted the sunny green field as temptation and the Warner brothers as a "warning" not to indulge his homosexual tendencies. The warning, he felt, came from his father—but he had never seen the house before and had no associations to it.

Ehrenwald was not surprised that Ronald did not recognize the house, as it was an almost exact description of a summer house he himself had viewed the day Ronald had the dream.

When Ronald later made a drawing of his dream at Ehrenwald's request, it was immediately recognized by the doctor's friends who had gone to see the house with him. The owner had insisted that the grassy stretch at the side should not be used, as it belonged to the next house. No objections would change his mind and so Ehrenwald reluctantly accepted these terms and paid a deposit. On second thoughts, he decided not to take the house after all, but the owner refused to return the deposit. Ehrenwald was angry with himself for giving in so easily and concluded that this was the reason for Ronald's mind reading; feeling decidedly one-down in the analytic relationship, he was catching his analyst out in a (less severe) unresolved problem of exactly the same kind as his own—impotence in the face of an authoritative father figure.

Dr. George Devereux, who collected together a number of papers describing similar cases and published them in a book, *Psychoanalysis and the Occult,* in 1953, was so struck by the fact that patients seem to spy on their analysts that he described their telepathy as an aggressive act, an attempt to impress, overcome, defeat, and castrate. Having suffered for three years from total frustration in psychoanalysis, I consider this statement itself very aggressive, blandly ignoring the possibility that the patient's desire for more open and equal contact with his analyst might possibly indicate something wrong with orthodox psychoanalytic technique.

However, the observation that telepathy seems like prying or spying is interesting, for it makes the "receiver" of the information the active agent reading the other person's mind, in contrast to the model of "mental radio" on which most telepathy experiments are based. The mental radio model implies someone actively trying to send out messages to a relatively passive recipient, whereas the alternative model, which immediately sprang to my mind after reading about telepathy in the consulting room, was that of *radar* scanning the environment for something either threatening or desired. This is congruous with the mind's "normal" ability to pick up subtle impressions and vibes without our being aware of them until they turn up in dreams. The radar model also puts telepathy in the same category as clairvoyance and precognition, which do not involve a sender.

The radar model does not, of course, explain *how* ESP is able to work in defiance of the present laws of physics, but it may direct our thinking and research into new areas—and Einstein himself once said that in physics the proper formulation of a problem is often more important than its solution. When I came to apply the psychic radar model to the dream diaries available to me, a pattern began to emerge which enabled me to formulate a hypothesis about the psychodynamics of ESP in dreams, and I can best introduce it by describing one of John's most spectacular displays of apparent ESP dreaming.

When I was enjoying the first success of *Dream Power* in the United States, John dreamed of writing a novel in which he himself was the hero and the villain was the head of a secret British government research station who intended to seize power for himself with the aid of a newly developed spy plane. In the dream John modeled one character of his story, a member of the government to whom he turned for help, on a real-life member of the British House of Lords, Lord Snow, but realizing he would have to give him a fictitious name, decided to call him Artie Shaw. The novel ended with John denouncing the villain and being covered with glory.

We had little time to work on the dream, for I was currently appearing on several TV and radio shows every day, but John felt its essential meaning lay in the spy plane. He confessed to feeling jealous of my success and resonated to the notion that one part of himself would like to have a magic spy plane that would enable him to keep an eye on me. Lord Snow was someone who had helped him in his career in England, but he had practically no associations to Artie Shaw, whom he had never met and remembered only as a bandleader of the 1930s and 1940s. We then forgot about the dream until two weeks later when I found myself at the studios of the Mike Douglas show in Philadelphia, being introduced to Artie Shaw, who was to be the principal guest that day.

This came as a complete surprise, since I had been told that the main guest was to be Peter Ustinov. Apparently he had cancelled at the last minute, and Artie Shaw had agreed to take his place. Naturally we told Artie that John had been dreaming about him, and when he expressed interest, we took the dream out of the folder we carried around with us and showed it to him. As he read the dream his face paled, and when he had finished he asked John to swear that he knew nothing about his private life. John replied that as a Britisher, he had not even kept up with his public life and had to confess that Artie Shaw was just a name to him. Artie then told us that he had once written a novel under the pseudonym of Adam *Snow,* but very few people knew of this. Within minutes the story of John's amazing precognition went around the studio, and when Mike Douglas interviewed me on the show itself, his first question was, "I hear your husband has been dreaming about Artie. Tell us about it." I did, and it stole the show. The rest of the guests, including Artie, joined in the discussion—and John was the hero of the hour.

There is a clear parallel here with the cases of ESP in the consulting room, for John's heart had a need which his head forbade him to gratify directly. Any attempt to upstage me deliberately would have made me angry and spoiled his own self-image as the good, sensible, unselfish husband—just as Freud's "Mr. Foresight" could not demand more attention outright without losing his self-image as the "good boy" patient and being told by Freud that he was showing infantile dependency needs. In fact, in both cases, the apparent ESP looks like a typical sly underdoggy maneuver, but whereas the psychoanalytic patients seemed to restrict themselves to spying on their analysts to impress them and capture their attention, John's psychic radar (or spy plane) swept out into the future to find a situation where he could upstage me on a major national TV show without taking any responsibility for doing so, and in a way that neither I nor anyone else could possibly grudge him. There was probably no other way in which John's underdog could have got what he was after so effectively, for Artie turned out to be interested in dreams, ESP, group therapy, and related subjects, and had a radio show of his own in New York on which he interviewed people working in different branches of the human potential movement. Having impressed Artie with his powers of precognition, there was every likelihood of John's being invited on his show.

The same pattern emerged in another of John's dreams some time later, shortly after I had agreed to fly to Canada for an interview with Kreskin, the mentalist whose amazing feats of apparent telepathy had intrigued me for a long time. John had hinted that it was a long way to go for half an hour's show, that I had a new book to write, and that it would be uncomfortably cold—all to no avail. Then he dreamed that one of his former industrial colleagues in the north of England, Dr. Y, had invited him to fly up as a distinguished visiting speaker to the company's local staff club. The meaning of this seemed obvious—John was being made aware of the fact that *he* wanted to be the one invited to the frozen north as a distinguished guest. The only puzzling thing was that the dream ended with John back at the London headquarters in the department where he used to work, talking to Dr. Y, who was sitting at the desk of the man who had been John's superior, Mr. H.

A week or so later, John received a letter from Mr. H with the totally unexpected news of his early retirement, and although John speculated about possible successors, it never occurred to him to think of Dr. Y, whom he took for granted was firmly settled in the north of England. Then a few days before I was due to fly to Canada, John received another letter from Mr. H with the news that Dr. Y was replacing him as head of the department. John was genuinely astounded, and I, of course, was very impressed. (For the record, John had never dreamed of Dr. Y before and had had no contact with him at all for several years.)

Again, John seems to have been exercising ESP in a typically underdoggy fashion, and this time the resemblance to "Mr. Foresight" was even closer, since John's dream revealed his anxiety that I should be so impressed with Kreskin ("the man from the north") that he might overshadow (become "superior" to) John in my affections. Since John could not protest this openly without losing his self-image of the cool, sensible, helpful chap, his underdog gave a demonstration of its extrasensory powers, saying to me in effect, "Look, I'm better than your old Kreskin. If you must go up and see him, the least you can do is call him up and say I should go too!"

Further examination of John's dream diary showed that he had two other dreams on the night he dreamed of Dr. Y, both of which repeated the theme of jealousy and anxiety. In the first I had been invited to Canada to *preach* and John was asking me to try to arrange for him to go too as part of the same series, but I refused. His heart felt clearly that in moving into the realm of ESP with Kreskin, I was invading a territory he considered his own, like preaching, and he was very resentful about this. The next dream produced a sharp reproach from topdog for even "dreaming" of such a thing. In it I was complaining about John's noisy snoring and said irritably that Kenneth Clark had never snored like that. By depicting Lord

Clark, the famous art critic who is author of the book and TV series called *Civilisation,* as my former lover, John was telling himself that I would lose all esteem for him if he continued to behave in such an "uncivilized" manner by grumbling openly about this issue. So underdog had drawn on his hidden talents and produced a splendid feat of ESP, thereby getting what he wanted in a truly "civilized" manner!

The same basic pattern emerges in the case of a friend of ours, Elaine, who has been keeping a dream diary for many years and has given us several examples of apparent ESP coming into play in a defensive role, as an extension of the heart's normal watchdog mechanism. In a dream that took place some years ago she saw her husband placing a gold necklace around the neck of a young, slim, dark girl and woke up in distress. It later transpired that her husband had indeed been having an affair with a girl of this description at the time and that on the very night Elaine dreamed of the gold necklace, he had given her one for her birthday that corresponded in detail with Elaine's recorded description of the dream necklace. While Elaine might have picked up subliminal vibes from her husband's behavior about the ongoing affair, there was no way she could have known about the birthday or the necklace, since her husband had bought it while away from home and destroyed the receipt. This seemed like a genuine case of ESP being used like psychic radar to spy on her husband. She said that he had never given her a necklace and had always seemed totally uninterested in birthdays, often forgetting hers altogether.

Soon afterward Elaine divorced her husband and became friendly with a man she considered marrying, but a series of dreams warned her against it. In one he was trying to kill her; in another she felt his cold shoulder on her back; and in another she dreamed of him without a penis. It later was revealed that he was trying to grow out of lifelong homosexuality by establishing a relationship with a woman, and Elaine's heart had obviously picked up vibes that he would eventually give her the "cold shoulder." None of these dreams need be explained in terms of ESP, but they all show her heart acting as a watchdog to protect her from the hurt of rejection.

In her next relationship, with a married man who insisted he was trying to get a divorce from his wife to marry her, she dreamed one night after he had been home to visit his family that she had been caught shoplifting. She protested her innocence, since she was not aware of having stolen anything, but the store detective assured her that she had been caught in the act by TV cameras hidden in portrait photographs on the wall. When he opened her bag, she was appalled to discover a bar of soap and a gold-topped bottle of scent which she had certainly not paid for. On waking, she had no difficulty in associating to the photographs, for her friend had returned from his visit home with photographs of his son, and she had suspected that he had arranged to have them taken as a birthday gift for the boy's mother. This thought hurt her deeply, for if it were true it meant that he was still playing the family game with his wife, which made a divorce seem very distant. Elaine had not expressed her fears because she felt she was not entitled to invade the privacy of his family relationship, but the dream made her decide to bring them out. It then transpired that her fears about the photographs were quite false but that her friend had in fact given his wife a bar of soap and a gold-topped bottle of scent for her birthday, gifts he had bought on the way to his home. Once again, as with the other birthday, it seemed that her resolve not to pry into her friend's dealings with his wife were overriden by her underdog heart, which "stole" the image of the soap and scent from his mind without her conscious awareness. On both occasions it seemed that underdog was using ESP to express a feeling censored by topdog that the gifts should have been presented to her and not to some other woman.

A similar case was reported by Dr. Emilio Servadio at a Conference on Spontaneous Phenomena in Cambridge in the 1960s.* He told the story of a young girl, Luisa, who dreamed that the mother of her fiancé, Guido, had on her finger a silver ring which had strange signs like hieroglyphics on the surface. The ring could be opened and she thought it contained a scent. On waking, she related the dream to her mother and then called Guido to tell him about it. He became very excited and said that he had just returned to Rome from Milan, where he had bought her a pair of earrings and his mother a silver ring which had a surface that could be opened and had strange writings engraved on it. Dr. Servadio interprets the dream psychoanalytically in some detail, concluding that Luisa was wanting by her spying to communicate her jealousy of Guido's mother and her resentment at not yet having received an engagement ring. As any open statement of these feelings would have been unacceptable, she resorted to the paranormal, which enabled her to express her needs without having to take responsibility for doing so.

My own extrasensory dreams in the past used to show a similar pattern of psychic radar used in a defensive manner, but in my case the defense was against threats of devaluation or hostile criticism. To my underdog, the worst threat life had to offer was any situation that could bring out the voice of topdog saying triumphantly, "There you are, I told you they'd find you out sooner or later for the monster/worthless slob /ineffectual idiot you really are—die, dog!" This is because I have normally tended to behave as the "rebellious Child" (in the terminology of Transactional Analysis), going out to get what I want in defiance of any authority that tries to restrict

*Described by Martin Ebon in *The Dream and Human Societies,* edited by G. E. Von Grunebaum and Roger Caillois. University of California Press, 1966.

or repress me—in contrast to John, who has tended on the whole to play the role of "adapted" or "conformist Child," giving in to external and internal authority and relying on sly, underdoggy maneuvers to get his needs met in ways for which he could not be blamed. The difference is probably due to the fact that my parents' demands in my early years were quite impossible (like the double-bind "Be detached—why don't you love me?), whereas John's parents were more reasonable.

The price paid by the rebellious child, as Perls points out, is guilt and constant fear of the slightest failure, which will enable topdog to proclaim that underdog's true colors are now exposed. So my heart was almost wholly occupied with defense, especially when I was succeeding or being accepted in any way. If I picked up any sign of rejection or vibes of hostility, my underdog would leap up whining about how misunderstood he was or trying to find grounds for rejecting the rejector, saying in effect, "I'm O.K.—it's you that's not O.K." The resulting fuss about how the world was ill-treating me provided a smoke screen against topdog and, as a bonus, would earn me some Brownie points with God, because God loves those who are persecuted and reviled. (This was an extension of my "victim" neurosis described in Chapter 8.)

I have had a great many dreams over the years in which this pattern is revealed, and several of them show underdog resorting to ESP when the potentially devaluing situation was not available to normal perception. The details are too numerous and too complicated to recount here, but I mention them because of the interesting fact that while most have been scattered, a whole bunch of them happened together in the few days preceding the resolution of my "victim" neurosis. Some were precognitions of events that did not occur until several weeks later. I was reminded of Jung's idea that ESP should perhaps be considered as "synchronicity"—the coming together of a whole series of meaningful coincidences at critical points in a person's psychological growth, brought about by an "archetype" beyond space and time pushing toward the surface of consciousness and "constellating" inner and outer events to reveal its pattern. I would express this in different terms by saying that whenever the self-torture game is coming to a head, and topdog is in danger of being overthrown, he renews his efforts at control, which forces underdog to thrash around wildly, using his psychic radar in all directions to get his needs met.

This casts new light on the common observation that a high proportion of authenticated reports of ESP outside the laboratory, both in dreams and in waking life, are associated with death, disaster, and other crises. The popular view that the operative principle in such cases is close emotional rapport simply does not fit the facts. Many of the reports concern people dreaming or having premonitions of disaster about slight acquaintances, when equally critical situations in the lives of their nearest and dearest go quite unheralded at other times. In my view, the operative factor in crisis ESP is some kind of underdog crisis in the life of the "receiver"—a particularly acute fear of topdog's threats of disaster or death. This causes underdog to mobilize his hidden paranormal energies and tune in to *any* appropriately disastrous event in the environment or even in the future, whereas on other occasions disasters go quite unnoticed. The notorious unpredictability and apparently erratic character of ESP springs from the fact that we are usually not in touch with our emotional crises, which go on below the surface of conscious awareness until something forces us to face them.

My model also explains why most of us never become aware of having exercised our paranormal faculties until some event shows that we have done so quite unconsciously. If ESP belongs to underdog, who is closely in touch with the body and its needs, then it is likely to be repressed by topdog along with all other "animal" and bodily energies. (Freud believed that ESP is probably an animal faculty which the human species has abandoned, and the idea that animals exercise ESP has been put forward by the distinguished British zoologist Sir Alister Hardy.) This would jibe with the widespread belief that primitive people, whose lives are closer to nature than ours, are also on much closer terms with the paranormal than we are. Nearer home is the fact that most of the people in our own personal experience who seem to be able to exercise ESP in a positive, predictable way are not ascetic "spiritual" types who deny the flesh, but worldly, bodily, energetic people who embrace the flesh and openly pursue the satisfaction of their natural needs.

Eileen Garrett, the famous medium, was notoriously outgoing, energetic, down-to-earth, and skeptical in her attitude to paranormal gifts, as is her British counterpart, the fun-loving Rosalind Heywood. Daniel Logan, author of *The Reluctant Prophet,* makes no secret of the fact that he loves to be the center of attraction, and developed his psychic powers in order to gain the attention and admiration he was denied as a child. Kreskin, who amazed me with his telepathic ability in a personal experiment on his show, describes in his autobiography, *The Amazing World of Kreskin,* how he developed his conjuring and genuinely psychic abilities alongside each other in childhood to impress his family and friends. He is a joy to be with, constantly bubbling with new ideas, and is well known for replenishing his energy with one large meal a day, which begins at breakfast time and ends at midnight! His life seems notably free of topdogs—which may be why I experienced "high" dreams for several days after being with him. And I am told that Edgar Cayce was no pale-faced prophet but a man who zestfully enjoyed the good things of life.

I believe that ESP, as we encounter it in everyday life, is a neurotic phenomenon, an aspect of ourselves that has gone under from childhood along with most of our natural joy and exuberance, because it lies outside topdog's framework of rules and regulations. In fact, our whole "normal" picture of the world as a space-time structure governed by cause-and-effect laws may be a topdog invention, a rigid artificially restricted framework into which we learn to organize our perceptions and thinking from earliest years—part of what Masters and Houston call the "cultural trance." Underneath, part of us knows perfectly well that reality is not like that; as has often been pointed out, there could be a constant background of paranormal perceptions hitting us all the time without our noticing them, for how would I know if the images in my mind are coming from the neighbor across the street or someone in the next town, except on quite exceptional occasions? (The success of some laboratory experiments on ESP may be based on this: the experimenter has a controlled situation where he knows what is being transmitted to whom and so can recognize it when it hits the mark.) Underdog normally takes great care not to do anything that will draw attention to the fact that his reality does not conform to topdog's rules —until he is pushed into a corner and driven to use his paranormal energies slyly and unconsciously, often hitting the mark in a haphazard and ineffectual manner.

This links with the well-documented observation that ESP seems to occur with high frequency under the influence of psychedelic drugs, for as Masters and Houston point out in their book *The Varieties of Psychedelic Experience,* the drug takes effect precisely by breaking the cultural trance. In Thomas Harris's terms, the Parent is "switched off" for a few hours during which the Natural Child emerges with an ecstatic upsurge of energy—and it is characteristic of the experience that time, space, and causality seem to be much subtler, complex, flexible things in this state of vision than under normal circumstances. There is also the joyful feeling of harmony with all living things that is the essence of mystical experience. When the doors of perception are cleansed, our ordinary everyday world becomes imbued with a sense of mystery and awe, and even the most mundane activities like walking down the street or eating a piece of cheese take on a new significance. As Wilson Van Dusen writes in *The Natural Depth in Man,*

> Suppose you had just this moment been born as a full-fledged adult, with your present mind and understanding. You would be absolutely stunned at the things and people around you. Most of the day would be taken up with "ohs" and "ahs" as you went around feeling things. It would be a frightfully impressive and awesome mystical experience. You would be stunned by the beauty of simple things such as the graceful form of plants. This is one of the hallmarks of the mystical experience, to find things fantastically beautiful and good *just as they are.* This comes before eating the apple of the tree of knowledge of good and evil. . . .

I have italicized four words in the quotation because to me they express the essence of what I am trying to say. By living in social structures where we are not allowed to accept ourselves as beautiful and good just as *we* are, we are condemned not to see anything else as it really is—which means that we are prevented from recognizing all those aspects of ourselves and the world that transcend the mechanical space-time framework. This is what all the world's great religions are getting at when they speak of man as fallen into illusion and servitude, and the Biblical story that links this with *knowledge of good and evil* shows great psychological insight. It is as if something has gone wrong with one of our highest faculties —self-consciousness; instead of using it to bring a new level of creative joy to animal life and the knowledge that we are made in the image of God, we have allowed it to pass over into moralism, which means that we delegate our powers to topdog systems of abstract rules. The Angel of Light within us falls by trying to control, instead of being a servant who guides spontaneity and creativity according to the demands of the situation.

Of those who think they do not suffer from this disease, I would ask, who gave you your present standards of right and wrong, proper and improper, sane and crazy? Even those who have discarded parental values have probably adopted such substitute parents as family, job, political party, church, nation, and spiritual leaders. In fact, said Blake, when we act according to what we think Christ would do or God wants, then we are worshiping Satan under the names of Jesus and Jehovah—and whenever we do that, we lose the power and dominion that is the birthright of our species. The odd flashes we get of that dominion, we call "paranormal"—but if we can learn to overthrow the reign of topdog and restore the Angel of Light to his proper role as servant, we shall find the natural child in us gradually growing up and training his powers in a world where transcendence of space and time is a perfectly normal aspect of the body-soul energy which Blake called eternal delight.

SHARED DREAMSCAPES

The psychic radar hypothesis explains why people in close emotional rapport sometimes dream the same thing on the same night, showing all the stages of ESP I have been describing. One of our students, Brad, recorded the following dream:

> Lynda and I are running through corridors and down stairways. I vaguely sense that there are other people running with me. We are running away from something, and there is a

definite sense of menace. We break out of a high, gray stone building into a field which is a maze of mines, trip-wires, and radar detection devices. I say to myself, "I'll never be able to run through all that!"—and at that point I fly away, thereby escaping.

On the same night, Lynda, Brad's friend, dreamed:

Brad and I are running through hallways inside this great big stone building and there's something after us. We run through this huge door into a field which is full of mines and radar detection devices. All of a sudden, Brad disappears and I start running, but I get caught by "them," whoever "they" are.

Brad added the following notes and interpretation of his dream:

I asked Lynda to draw a diagram of the field and building while I did the same. At no time in our recounting of the dream did either of us mention a road or trees. Yet our drawings [which he enclosed] each have a building, trees, a field and a road. The diagrams are exact mirror images of the dream scene.

These dreams took place on the same night and, as far as I can determine, at the same time of the morning.

It's almost embarrassing to say it, but a Freudian interpretation may be valid here. The castle, I feel, is a penis, or in a larger sense, sex. Lynda and I are menaced by it and run. I leave her. This is a fairly accurate account of Lynda's and my initial encounters. The specter of sex raised its head, and we both ran for our own reasons, I being the one who left Lynda.

Adding my own note, I would say that Brad's flying represented his feeling that there was only one way to escape from sexual involvement, namely "rising above" it with spiritual disciplines such as yoga, meditation, and so on, which he practiced regularly. In fact, the dream throws an interesting light on his reasons for doing these disciplines. It is impossible to say for certain from the dreams who tuned in to whom that night, but I suspect it was Lynda who remained emotionally involved while Brad escaped. Her internal radar may have tuned in to Brad's dream in a desperate effort to get closer to him and to find out how he really felt inside about the relationship. Her dream gave her the truth—from her viewpoint, he had vanished.

John and I had a similar experience but with a significantly different twist. I dreamed I was staggering over a bridge with dishes of food I had cooked for a party, feeling very resentful because no one stopped to help me. On the same night John dreamed that I had agreed to prepare food for a bridge game. I associated my bridge with the book I had been reading before falling asleep—Baba Ram Dass's *Be Here Now,* in which there is a picture of someone standing on a bridge watching his reflection in the water, with the caption: "You are standing on a bridge watching yourself go by: Wow! Look at that!" So I looked at the dream in terms of "This is how I see myself at the moment—overburdened with people's requests for lectures, interviews, dream interpretations, and so on," which is how I interpreted the food (food for thought). I had, in fact, complained bitterly to John about this the previous day and had insisted that he cancel all my forthcoming engagements.

He interpreted his dream in the same way—as his feeling that I had allowed myself to prepare too much food for thought (he was aware of my use of this symbol). Neither of us even noticed the odd detail of the "bridge game." We don't play bridge, and as far as we can recall, no one mentioned bridge the previous day; nor had either of us dreamed of a bridge game previously. It was only much later, when we were working on my "victim neurosis" and beginning to realize that I almost deliberately sought suffering and rejection to get God's love, that John remembered our bridge dreams. On reading through his own, he concluded that his heart had pried into my dream that night and seen me staggering across the bridge with all the food—and realized that it was all a big *game* on my part. Had we understood his dream at the time, my victim neurosis might have been resolved much sooner. Obviously John's heart had picked up something phoney during the day but he probably knew that if he had said so or even dreamed it directly, I would deny the whole thing and be angry. ESP, however, would be a different matter altogether—the very fact that he caught me out in my own dream game would be enough to drive the point home and probably make me laugh too.

Was John's prying selfish or unselfish? Was his psychic radar activated by resentment of my "game" and the trouble it caused him, or by concern for my welfare? Probably both. The distinction between selfishness and unselfishness is a topdog invention that does not apply at the deepest level of the personality, where concern for others flows naturally from healthy self-love. Moreover, since we have been working together to overthrow our topdogs, John and I have had several experiences in which ESP seems to have enabled us to share knowledge in dreams. A striking example occurred one night after John had finally resolved, through intensive dream work, a problem that had bugged him for most of his life. He was sleeping on the open porch that night and dreamed of an Indian servant coming in just as light was dawning, chattering loudly as she prepared to clean up the house. (It was actually 3.00 A.M. and totally dark outside.) John told her to be quiet as I was still asleep. Then it suddenly "dawned" on him that we don't have a servant, and he instantly became lucid, aware that he must be dreaming. Immediately the light became almost supernaturally intense and the landscape outside took on a golden hue which reminded him of Van Gogh country in the south of France. Space seemed expanded and deeper, just as it does under psychedelic drugs, and John noticed figures dressed in old-style peasant costume walking in slow motion across the fields. He recognized one as a friend he had not seen for a long time, except that she was now a very old woman whereas in real life she would be in her early thirties. As she passed him, she looked up and smiled knowingly, as much as to say, "Now you *know* what it's really like." John's excitement woke him up, and his first words to me were "I've joined the club. I've had my first high lucid dream. Now I know what you've been talking about. We share the same phenomenal world."

I then told him my dream—of sitting in an apartment when the dividing wall crumbled so that two apartments became one. I linked my dream with the fact that in the past I had always been very defensive about ESP and had agreed with Theodore Roszak when he had said a few years earlier that if people thought the problem through, they wouldn't really want ESP at all. Imagine, he said, picking up every little thought and vibe, and having yours picked up too. What a way to live! My feeling had been corroborated when I put a request to dream power, some time later, asking how I really felt about ESP. I dreamed of being in a very dangerous place on an island, surrounded by snakes and animals, and making myself a kind of nest in a hole in the ground to protect myself from danger while I slept. I was very comfortable there and was reminded of Don Juan's "place of power" where magical things can happen. Then I was disturbed by someone who insisted on sharing my place and I felt very angry at the intrusion. Then the scene changed and I was back in my childhood home, trying in vain to go to sleep but constantly being disturbed by voices and ringing telephones. The dream made it clear that I saw ESP as a possible weapon of intrusion into the private recesses of my being, my secret haven from the nagging presence of topdogs. I later discovered that Dr. Jule Eisenbud found telephoning to be a common symbol for telepathic communication.

The fact that I allowed the dividing walls to crumble on the night John experienced his high lucid dream is interesting because it corroborates my hypothesis that the "high" state is one in which topdogs are switched off or absent. If that is the case, then I had nothing to fear and I no longer needed defenses. Had my dream continued, I might have found myself sharing John's phenomenal world literally by joining him in the golden landscape, an experience we have had on a few occasions since. (Had I realized earlier, of course, that ESP is the secret weapon of underdog rather than topdog, many of my misgivings about ESP would have disappeared.)

TOWARD FREEDOM: HIGH, LUCID, AND OUT-OF-THE-BODY EXPERIENCES

In high lucid dreams like the one I have just described, we get glimpses of what human consciousness can become when head and heart are together *in an experience of true freedom.* This consciousness seems completely to transcend the limitations of ordinary sense experience and show us a reality of many different mansions or dimensions, so that people who have experienced this kind of dream state have often spoken of leaving the ordinary body behind in sleep and going off in an altogether subtler body to explore other worlds. I know from personal experience what they mean, yet I believe it is important even at this stage not to forget the relationship of the dream to waking life. John's golden dream showed him in depth a vision of the glorious freedom beyond space and time which his heart had already known for at least a fleeting moment during the day, and my own dream experiences of the same type have increased from about one percent to ten percent of my total recorded dreams since I have been working on the resolution of topdog/underdog conflicts in waking life.

In fact, our dream diaries and those of many of our students show a definite pattern of growing enlightenment or "wakefulness" emerging in dreams as the dreamer struggles toward freedom from topdog's mental cages in his day-to-day living. When an insight accepted by the head from dream work eventually begins to penetrate the heart, the person will often have what we call a "self-reflection" dream, similar to my verification dream described elsewhere. As a result of my having worked on my "victim neurosis," the message was slowly but surely getting through to my heart that I didn't have to play the silly suffering game any more, and the first intimation came in the dream when it suddenly dawned on me that I didn't have to be a servant—and I immediately gave notice! Similarly, when John's heart finally got the message that freedom was more valuable than security, he dreamed of going miserably to work and suddenly realizing that he didn't have to be there. One of our students, who had been working on an "ugliness/worthlessness" image instilled into her by a parental topdog, dreamed that she looked into a mirror and was astonished to find herself beautiful (which she was), while another student realized in a dream that her mother was dead and could not restrict her any longer.

All these self-reflection dreams show head and heart coming together in a new clarity of vision which illumines waking and sleeping consciousness alike. The next stage seems to be the "pre-lucid" dream in which the dreamer debates within himself whether or not he is dreaming. Sometimes he will decide for certain that he is dreaming and pass over into a fully lucid state, while on other occasions he will be distracted by the events of the dream and fall back into an ordinary dream. Pre-lucid dreams are often associated with a "false awakening," an experience common to those who have lucid dreams. This is a dream in which we believe we have woken up, only to discover that we have not. Sometimes the realization that we are still asleep gives rise to a fully lucid dream, while on other occasions we discover the mistake only on real waking.

I experienced many pre-lucid dreams some years ago while I was studying with Joel Goldsmith, the American mystic and healer. While my head accepted that I was one with God and that all the powers of the universe were mine, my heart was somewhat slow in following suit. But gradually as I learned to confront my fears, I was often able to catch myself out in waking life as some catastrophic expectation threatened to overwhelm me—and this was often followed at night by a pre-lucid or lucid dream, usually of the nightmare variety. As the threatening image advanced, a trickle of consciousness would reach me and I would start debating, "If this is a dream, it has no power over me," and on some occasions I would feel confident enough to turn around and face it. In the early days my lucidity was not enough to sustain my strength, but later on I was able to overcome dream enemies by this method. These pre-lucid and lucid dreams reflected my growing realization in daily life that most of the things I feared were figments of my own conditioning and had no power to restrict my life.

A Buddhist friend of ours reports similar experiences after training in "mindfulness," which consists of being aware of one's body, emotions, and thoughts in waking life, a state of consciousness that should ideally carry over into sleep. In fact, for those pursuing religious disciplines of any kind, dreaming is the best way of mapping one's progress in order to know how much of what we believe is still only a head trip and how much of the teachings have reached the heart. The "false awakening" is usually an indication that we have not awakened as fully to something in our life as we believed. As one of our students wrote, "I remember one dream involving a persistent and threatening homosexual theme in which I must have thought I'd woken up four or five times. This in keeping with my tendency in normal consciousness to feel, 'Aha, now I've woken up, am together, have transcended trivial ego games'—at least thrice before chicken soup! Oh, well . . ."

The next step is the "ordinary" lucid dream in which we become aware that we are dreaming but do not experience the expansion of consciousness that characterizes a high lucid dream. Such ordinary lucidity is normally only partial in that we know we have power to change the dream, yet often decide to use it only within absurd limits of thinking, which again is a reflection of the waking state. For example, one of our students who was coming to the resolution of a sexual conflict reported that when she became sexually aroused in a lucid dream, she started searching for a mate and found herself in a downtown brothel—instead of conjuring up her hero in a desert tent, or whatever! On another occasion, she found a mate but could find nowhere to be with him—and it never occurred to her to *make* a place. These dreams show that she was not yet in control of her own energies, but her latest dream —which was her first experience of high lucidity—indicated the approach of resolution.

In the dream, she was informed that she could choose either to have intercourse in public with a fantastic dream lover and be strangled by him afterward, or never to have sex again. Her growing desire for a life lived to the full rather than a living death led her to choose the former, and as she was being led into the arena she suddenly became lucid. Instead of waking herself up or changing the scene, she decided to trick them all and go along with the game; and as she laughed to herself at how she would get up and walk away at the end, the environment expanded, the colors deepened, and she was high. Then the scene changed and she found herself flying in an extraordinarily high state, going through walls and windows without difficulty, and although she had been looking forward to the sex, now her deprivation did not seem to matter because she was enjoying other even more exhilarating experiences. The dream links with our finding that sexual arousal in a lucid dream can lead to flying and out-of-the-body experiences when the sexual energy seems to flow up through the whole body; whereas on other occasions, when the energy stays around the genital area, an ordinary but intense sex dream results. Perhaps there is some connection with the old tradition which holds that sexual energies can be transmuted into other potentials when not confined to the energy-dissipating sex games of prestige and role playing in our society.

I believe that a high lucid dream indicates the winning of a battle over topdog, probably the final battle of one particular conflict area, whereas the earlier stages of pre-lucidity and ordinary lucidity indicate steps on the way to liberation. Just as it is difficult to remain mindful of oneself and one's actions constantly in waking life, so it is difficult to retain lucidity in dreams. In my early experiences I would be so excited when lucidity came to me in a dream that I would jump or call out in excitement, which immediately caused me to awaken. I gradually managed to train myself to remain calm and in this way was able to explore the lucid environment. Even so, lucidity would come and go and I would find the dreamscape changing as in a normal dream. For this reason I was grateful to Carlos Castaneda for giving us, in *Journey to Ixtlan,* Don Juan's technique for stabilizing the lucid dreamscape. Don Juan suggests that we tell ourselves before falling asleep that we shall look at our hands in a dream and become aware that we are dreaming. We are then gently to shift our gaze to some object in the environment, and after a minute or two, bring the gaze back to our hands. This exercise should be repeated until the dreamscape is stabilized. "Every time you look at anything in your dreams, it changes shape," he told Carlos. "The trick in learning to *set up dreaming* is obviously not just to look at things but to sustain the sight of them. *Dreaming* is real when one has succeeded in bringing everything into focus. Then there is no difference between what you do when you sleep and what you do when you are not sleeping." Carlos reports that he had little success with the technique at the time of writing the book, which I would attribute to the fact that he had not yet gathered his powers together in waking life, as Don Juan continually reminded him. In my own experience, Don Juan's technique proved valuable *after* I had achieved some measure of lucidity in dreams through dream work in waking life.

Don Juan went on to teach Carlos how to travel in dreams—to pick any place like a friend's house, or a park, or a school—and go to it in sleep. This is what occult and psychic literature calls "out-of-the-body travel" or "astral projection." I agree with Celia Green when she writes in her book *Lucid Dreams* that it is not possible to discuss lucid dreams without considering their relation to out-of-the-body experiences, and Oliver Fox in his book *Astral Projection* distinguishes between a "higher" and "lower" grade of lucid experience, of which the former is out-of-the-body travel. This is similar to my own distinction between ordinary lucid dreams and high lucid dreams, for with the latter comes the feeling of a lighter body and the ability to fly and go through walls. Or, as Oliver Fox describes the experience, "With the realization of this fact [that I was dreaming], the quality of the dream changed in a manner very difficult to convey to one who has not had this experience. Instantly the vividness of life increased a hundred-fold. Never had sea and sky and trees shone with such glamorous beauty; even the commonplace houses seemed alive and mystically beautiful. Never had I felt so absolutely well, so clear-brained, so divinely powerful, so inexpressibly *free!* The sensation was exquisite beyond words. . . ."

Whether this kind of experience really indicates that consciousness is leaving the physical body behind in some way,

I do not know, but it certainly feels like it. The trouble is that it is often so contaminated by ordinary dream consciousness that we cannot tell. In my own work I have distinguished between two kinds of out-of-the-body experiences—those that occur in a dream environment and those that actually take place in my bedroom at home in which the dreamscape corresponds to my actual sleeping environment. Very often, I will find myself in the latter state, wandering around the house, but will find small details out of place, like a window in the wrong wall, a potted plant we don't possess, or a door where there isn't one. In one such dream I went into Fiona's bedroom and told her I was an astral body, at which she laughed and pinched me. At the time, she was staying with her grandmother seventy miles away! On another occasion I decided to conduct a "scientific" experiment by rearranging the furniture to prove to myself when I woke up that I really had been out-of-the-body. This, of course, was nonsense but it seemed quite sensible in the dream.

I am still very much in the exploratory stage with this kind of "dream,"* but I will mention one other strange experience. I had become lucid in a dream and had been flying around when I suddenly felt myself on the shoulders of a giant. All I could feel was the pressure on my legs as I struggled to free myself from his grasp. Suddenly there was a kind of click and I twisted free of my dream body in some way, to find myself in an even lighter body in a dreamscape of brilliant white light looking down on the Eternal City. (I don't really know what the Eternal City is, but that's what it was in the dream.) I wondered whether to fly over and explore but feared the light might burn me, so I continued my explorations in the new world of even more brilliant colors and strange shapes. I felt rather uncomfortable here, as it was more a world of patterns

*I should appreciate hearing from readers who have worked in this area.

and light than people and objects, and I wasn't sure how to function. Much later, on reading one of my old theosophical books, I was surprised to discover that it is a fairly common experience for those practiced in the art of astral travel to feel a dragging on the legs or in the body, as one dream body leaves another; also that the world of the higher astral is one of light and pattern.

My own inclination would be to interpret the experience in terms of topdog remnants still trying to prevent us from expanding our consciousness from their tightly constructed space-time world into other realms of freedom. Perhaps it is because John and I approach the exploration of these inner spaces by setting it firmly in the context of psychological and spiritual growth, instead of pushing the river with occult practices, that neither we nor our close students have ever had a bad experience in a lucid or out-of-the-body dream. I believe that the so-called spirits, entities, or thought forms that are reported to pursue the unprepared explorer are projections of unresolved conflicts, the topdogs and underdogs who also pursue us in ordinary nightmares. We can all win the dream game if we persevere, and the prize is freedom to explore new realms of experience without fear.

APPENDIX

Pages from Sara's Dream Diary and Glossary with an introduction by Sara

I am a thirty-five-year-old single professional woman living and working in Philadelphia. I am an only child on friendly terms with my parents, who live in a nearby planned community. My father is a top-level administrator who has little time away from the ceaseless demands of his job, and my mother is involved in numerous social and civic activities. I feel myself quite different from both parents and am frustrated by their conventional and seemingly insensitive attitudes on human issues I consider of utmost importance. Most baffling to me is their total indifference to things psychological. Dreams, for them, are nonsense. Groups are unthinkable. I have thus been very shocked to discover from my own dreams that I have incorporated into my own psyche their very attributes that have caused me the greatest pain. (My dream of Pope John [on page 272] is an example of this.) My struggle to find and claim my own identity is punctuated by alternating efforts to be like and different from my parents. Caught between these conflicting ambitions is my real self—the happy, adventurous, playful, and loving underdog of my personality who makes himself known very forcibly in some of the following dreams.

I have always been interested in dreams and was understandably upset by the first dream in the glossary. I call it my "alcoholic" dream, and it was this that forced me to seek help in Ann's dream group. To understand the whole problem one has first to know that I have a serious overweight problem, and like many other people with this problem, I am a frequent though unsuccessful dieter. Not long before the "alcoholic" dream friends of mine who were in fact members of Alcoholics Anonymous suggested that I join with them in a group which would try to apply the principles of the AA program to weight reduction. The group didn't come off, but I began to read AA literature to see if I could apply the program on my own.

Recognizing a great emotional similarity between myself as an overeater and the alcoholic, I began to worry about my own drinking habits. From a family of teetotalers and moderate drinkers, I had no basis for determining if I had a drinking problem. However, feeling emotionally kin to the alcoholic and knowing that I drank more than my family would have considered proper, occasionally getting quite high, and finally discovering that drinking made dieting impossible, I began to experience considerable anxiety and confusion about what my problems were. My dreams helped me solve my "drinking problem" by showing me that the part of my personality responsible for my overindulgence is my irrepressibly high-spirited underdog who refuses to be "controlled" and put down. By listening to his needs and allowing him expression in various ways, I am now able to take a drink whenever I feel like it without anxiety and abstain when I feel like it. In Perls's terms, my dreams have helped me learn to function according to the situation and not to a rule derived from an inner topdog and his unrealistic catastrophic expectations.

I wish I could report similar success with my weight problem, but I still fluctuate between rigid dieting and overeating. I can take off and put on weight faster than anyone I know. My problem may have something to do with the fact that I was a premature baby weighing less than three pounds at birth, and during my first year was so stuffed with food that I probably got the message that to eat and get fatter is to be approved of and loved. However, my dreams are now showing me in no uncertain terms that my overweight is only a symptom of a deeper conflict, as was my drinking. One thing I know for certain—underdog will continue to cause one problem after another until I learn to accept him as he is. I am sure that as I care for underdog, my weight will take care of itself, for overeating is not underdog's natural need.

Space limits the number of dreams that can be reproduced here, and the few I have chosen illustrate two major points—how certain symbols recur over a series of dreams, and how dreams can radically change our lives in very down-to-earth ways by showing the destructive conflicts that determine our behavior. I have many more dreams dealing with other issues in my life, and as I gradually free myself of parental and societal topdogs, I expect to see a resolution of many emotional, religious, and spiritual conflicts. In the process many cherished values may have to go, as my last dream made clear. In it I was trapped in the doors of a cathedral and the only way I could be rescued was to destroy the cathedral. I am sad about this, as it had seemed to me to possess a kind of beauty—but it is no permanent home for underdog. I am sure that I will eventually find true spiritual treasure among the rubble—but that's another story. In the meantime, I have some comments about the way in which I work with my dreams.

I am fortunate in having generally excellent dream recall. It is the rare morning that I wake with no dream in mind. On my return from the retreat in the Pocono mountains where I first met Ann and John, I immediately began compiling a dream diary. At first I did very little work with my dreams, but somehow felt good and more in touch with myself just from writing the dreams down.

As I began actually to work with my dreams, I found a mistake that I had been making. A day's events not recorded with the dream usually couldn't be recalled and that dream would remain forever a mystery.

In working with my dreams I first use associations for any clues that come from that method. I'm seeking to determine what event, problem, or feeling the dream is dealing with. I then look at the dream as a picture, which helps me to understand the inner action. Finally, with dream elements that remain obscure or with ones that feel important, I engage in dialogue or, less frequently, use a monologue.

With either monologue or dialogue I use a combination of speaking out loud, sometimes screaming or crying, and writing. The writing of the dream character's statements I find to be very important. Often in reading the dialogue I see things that I didn't hear myself say. Sometimes I use a tape recorder, thereby catching my tone of voice, significant pauses, and other subtle clues that help me identify topdog and underdog characters.

The glossary is a great innovation, fairly easy to do, and a tremendous help as symbols really do begin to repeat themselves. The recurring symbol is fun. I get a particular charge when I understand a symbol in a new dream that I missed in an old one and then can decipher the first as well. Either way, it's like two dreams for the price of one.

With the wealth of dreams that I am now able to record, I find I can be selective as to the ones I give the full treatment. Usually I have at least one really good dream per week. By really good, I mean vivid, dramatic, interesting ones. I try to be alert for dreams following days on which significant events occur in order to savor fully all my feelings and reactions to whatever has happened. Times of decision also provide helpful dreams. I have only recently gotten the hang of putting questions to dream power and lately used this technique with startling results.

Naturally, dreams have provided me with many valuable insights into myself, but most important to me is that slowly and gradually I am learning to trust my dreams for guidance as to my own needs and the things that are right for me. A slow learner in this area, I find that dreams are persistent and will create story after story to get their message across. On critical matters that I ignore or on which I am making a wrong decision, dreams begin to show me the consequences of failing to heed their message. This type of dream, sometimes a nightmare, is not pleasant, but very useful. Having never suffered from recurring nightmares, which may be another matter, I am convinced that it is my own blindness to dreams' messages that causes my nightmares. I think nightmares are a last resort for dream power.

The following pages illustrate how I keep my dream diary, though I have amplified each dream's explanation with notes for the benefit of readers. Whenever a symbol recurs, I transfer it to a permanent dream glossary at the end of my loose-leaf diary, where I also keep a record of my most important transforming symbols.

5th July, 1972 *1*

Dream: I am sitting at a table in a café with Elsie. A friend of mine, Margaret, comes in looking awful and starts talking about all the trouble she is having, and how she is afraid she's an alcoholic. She continues talking about how awful alcohol is, what a problem, and how she will have to quit. I am unable to say anything but simple mouthings like "that's too bad." I keep trying to tell her that I'm an alcoholic too and that she can get help from Alcoholics Anonymous. But I cannot bring myself to admit in front of Elsie that I'm an alcoholic. I feel absolutely terrible about not saying anything. She doesn't even know about AA and I can't tell her the organization even exists.

Day's events and associations: I had entertained my parents for dinner the previous evening. I had served them drinks but taken none myself. In fact, I hadn't had a drink for about a week.

Dream theme: Inability to speak.

Puns: None.

Dream symbols:
Elsie — an acquaintance—judgmental, religious, proper, naïve (reminds me of mother).
Margaret — a friend—delightful, interesting, energetic, nonjudgmental.
Inability to speak — can't share with parents.

Notes: As I haven't seen either Elsie or Margaret for some time, they must symbolize someone else in my life or parts of myself. As Elsie reminds me of my mother, I guess she stands for my parents, with whom I would like to share my problems but can't. I've never talked to them about my drinking problem, and as I've never overindulged in their presence, they know nothing about it. I don't want to tell them because, while their heads know that alcoholism is a sickness, their hearts still see it as lack of moral fiber—and I just couldn't bear them to think of me like that. On a subjective level, Elsie also represents my own judgmental topdog (derived from parents), and I just can't bring myself to admit that I need help. The dream also makes it quite clear that it is the easygoing, energetic, nonjudgmental side of myself (Margaret) who has the booze problem.

Dream message: I feel that I do need help badly over my "drinking problem," but can't admit it because this would mean losing my self-image of the independent, firm, strong-willed Sara. I am also in two minds as to whether or not I really am an alcoholic, though I'm fairly sure I'm not, and I wonder whether the AA program is for me, though once again, my heart seems to think not.

Action: The dream shook me so much that when Mary called the following morning and told me that the Faraday dream workshop had been rescheduled for the 7th, I called up immediately and made arrangements to attend. This was the first step toward helping myself.

6th July, 1972 *2*

Dream: The setting is a tent which is dark, candle-lit and hot —a kind of Arabian Nights atmosphere. In the tent is a chiropodist, a man with dark, curly hair wearing a dirty white T-shirt, tennis shoes, and no socks, and a dark-haired woman. Somehow the thickness and tread of the man's tennis shoes is important. He says he is learning to make shoes for people who need their shoes built in special ways because of foot defects. We discuss shoes for a bit, mostly with this man describing ways they can be built up inside to do one thing or another. Then a bearded man comes to our tent and asks the girl what she does. She says, "I'm a surgical nurse," and he replies, "Good, you'll do."

The scene shifts and the tent is now set up as an operating room, and the person on the table is under a sheet having his appendix out. The operating room is lit by candles and is relatively dark. The young chiropodist is doing the surgery with great anxiety, and the nurse is assisting. The bearded man is in the background, sort of in charge.

Day's events and associations: My friends persuaded me to attend an AA meeting the previous evening and I went along out of curiosity. It was held in the basement of a very old church with poor lighting. I hadn't seen my chiropodist for weeks, but when I first visited him about a year ago, he told me that my feet were too small for my frame and prescribed special shoes for me.

Dream theme: I watch a surgical operation.

Puns:
a. The chiropodist has a firm, sturdy "soul" (sole) in contact with the earth.
b. He talks of "building up the soul from inside."
c. There is a "cutting out operation."
d. The chiropodist is "working in the dark."

Dream symbols:
Arabian Nights — dark, mysterious, strange, exciting.
Chiropodist — my real chiropodist, who also represents the healer within me.
Tennis shoes — make the feet look big; large soles in contact with the earth.
Sole — soul: the basis of personality and that which must be in contact with the earth.
Bearded man — George X from AA.
Appendix — useless organ that can cause a lot of trouble.
Operation — cut out my useless drinking.
Operating in dark — feeling of confusion about what I'm doing, and what I should or should not do about my drinking.
Body on table — myself.

Notes: I really like and trust my chiropodist and find him a "healing" person. As I haven't seen him for some time, he must represent the healer within myself who is very anxious about cutting out booze along the drastic lines of AA. In my heart I know that my drinking is a symptom of a deeper unhappiness and that real healing lies in building up my soul "from within"—getting in touch with my own inner resources, building up my self-confidence, enlarging my soul so that it stands firmly in contact with the earth. Cutting out booze is like removing a symptom, which may make me feel better superficially, but does nothing toward the cure of my soul.

Dream message: Despite my misgivings about cutting out drink and in the full understanding that it will not solve my problems, the dream indicates that it is healing for me to do so *at the moment* as an emergency measure. However, there is also hope of a more permanent inner healing.

Action: I did not take any drink—and I hope for help in depth at the dream weekend.

9th July, 1972 *4*
(last night of weekend dream group)

Dream: There is a small elephant playing in the bathroom. He is filling his trunk with water and spraying it around the room. After a time, he climbs into the tub, which is filled with water, and as he is playing there, he slips and falls. In falling, he hurts his trunk slightly—there is a quarter teaspoon of blood on it. I come in and pick the elephant up out of the tub and carefully dry him off. Then I take a tissue and gently pat the blood from his trunk.

The scene changes, and we are now in a filling station. It is pouring down rain. I am in the back seat of the car. Ann Faraday is outside the car helping me to get the elephant into the car with me. I try at first to push the elephant a little away, saying "nice elephant, nice elephant," but having no success, give up and allow him to kiss me. It is a happy reunion.

Day's events and associations: I had worked on Dream 1, above, and another one in the dream group, with very positive results. Using the Gestalt method of dialogue, I had told myself with the voice of a wise old man I know, "You're a good person; take it easy; dig down, find the truth; you don't have to do it all at once. You have time."

Dream theme: Making friends with a playful animal.

Puns: a. Slip — a "slip" from abstinence.
b. Fall — a "fall" from virtue and grace.

c. Pick up — rescue from the fall.
d. Dry — teetotal.
e. Wet — drinking.
f. Taking a back seat
Being a passenger — relinquishing "driving" role.

Dream symbols:
Elephant — my overweight self: aspect of my animal nature that overeats and overdrinks.
Bathtub — the last slip I had prior to the dream had involved drinking gin; the setting, I think, involves the pun of bathtub gin popular in the days of prohibition.
Quarter teaspoon blood — smallest possible injury.
Filling station — place of refueling—which I was already feeling in terms of revitalization through getting in touch with my dreams.
Ann Faraday — herself.

Notes: I feel that taking a back seat is a matter of practicality, for, after all, an elephant in the front seat waving its trunk in front of the driver would be quite a hazard. I must, for the moment, relinquish my "driving" role if I am really to pay attention to underdog. My ambivalence about the elephant is not that I don't like or want it in the car with me, but I just don't like being slobbered over. But given a moment to get used to the idea, I let him kiss me and feel very good about it.

Dream message: My slips and falls from abstinence have not seriously hurt me, and each time I have been able to pick myself up, dry myself off, etc. The dream gives a vivid picture of that part of my personality involved with overeating and overdrinking—the young, playful, loving aspect of myself. I am regaining much internal energy from working with my dreams and am trying to go along with Ann when she encourages me to accept myself as I am *now*—not when I'm thinner, not when I'm nondrinking—but just as I am now at this moment. I fear that my own affectionate nature will be a bit overwhelming if I allow it expression, but my dream says I might even enjoy it once I get used to it! The dream augurs a happy reunion of top-Sara and underelephant, a creative synthesis of my former "drive" and newfound playfulness—resulting hopefully in a more relaxed and altogether less serious attitude to life.

Action: I went home with a transforming symbol—the elephant—to remind me that the overindulging, sloppy side of myself I so despise is also the respository of joy, energy, and love. If I try to "cut out" my overindulgence by any radical operation, I shall be in danger of destroying many positive qualities also, which means that it will almost certainly fail! My future course of action is to love the elephant, allow his needs expression, tolerate his slips, and help him grow to the point where he won't want to indulge. In the meantime, I shall forgive him every time he slips, instead of chastising him. After all, he's only a baby—and he's got me to look after him! I shall see that he comes to no harm.

This dream is a favorite. I love it so, because it shows me taking care of myself and accepting myself. As a child, a favorite story was Dumbo—the baby elephant who learned to fly when someone took the trouble to believe in him.

31st August, 1972 *8*

Dream: I am in a house that seems isolated—with land as far as the eye can see in every direction. It's night, very late. Someone enters from having made a dangerous mission across no-man's-land. He is followed by a puppy belonging to one of the girls in the house. The man is annoyed because the puppy, which is rather large, jumps up and nips a lot.

I then have to make the same journey. The man who has entered the room warns me about the puppy again and everyone says to be careful. I try to kick or throw a ball that it can chase and keep its distance, but am never able to get the ball more than a few feet from me.

Day's events and associations: I had taken a couple of drinks the previous day, but nothing excessive. I had also seen my chiropodist. I made plans to purchase a tape recorder—a sort of investment in dream power.

Dream theme: Pushing away an animal.

Puns: Nip—to take a drink.

Dream symbols:
House — safe place.
Man — courage, determination (my father?).
No-man's-land — a dangerous place (the world?).
Puppy — free, mischievous, playful part of myself.

Notes: Although my mother takes an occasional drink herself, she considers two in the same day overindulgence—therefore I had sinned against my parental topdog by taking two drinks the previous day. My father used to arrive home late from work when I was a child, and I suppose I may have wanted to greet him rather effusively like the puppy, but never dared because I knew he wouldn't like it. This may indicate that it is the small, affectionate, boisterous child who drinks out of loneliness for her father even now.

Seeing my chiropodist (healer) and my intention to buy a tape recorder for my dreams were probably the triggers for the emergence of underpuppy, who smelled some attention in the offing.

Dream message: My present situation reminds me of the loneliness and isolation I felt at home as a child, with my mother not really being able to understand my problems and my father a virtual stranger to me. As I had not seen my father at the time of this dream, the dream probably represents my own serious, earnest, hard-working topdog, who insists that I face the big, hostile world with courage and determination, which means that I must control myself and quit "nipping." The dream shows me buying his catastrophic expectation that underdog is dangerous and must be put at a distance—but there is a secret saboteur who prevents the ball from going too far away. This must be underdog himself, who is determined not to be put down. Moreover, the dream shows clearly that his nips don't hurt me in any way.

Action: I told my parental topdogs to get the hell out of my life—and drank their health with a couple of gins! I bought the tape recorder the next day in the hope and expectation that my dreams will help me build up my soul from within, which means death to all topdogs. And who wants the world to be a no-man's-land?

Permanent dream glossary: The recurrence of an animal in my dream paved the way for the start of my permanent dream glossary, which appears at the end of this Appendix. A recurring symbol in dreams indicates that it has a special significance in the dreamer's psyche, and it is useful to keep a separate glossary of such symbols at the back of your dream diary to which you can refer when necessary.

7th January, 1973 *14*

Dream: I am inside a car with several other people. Nan comes up to the car—she is wearing a topless bathing suit and is deeply tanned. The bottom of the bathing suit is not bikini style, but more like shorts with an elastic waistband. Her breasts look rather strange—they come to such a point that they are almost triangular. She makes a remark about competing with me to make her breasts as large as mine. Meanwhile, I am thinking that if I have lost enough weight I may have time before September to go and get a tan in Puerto Rico.

Day's events and associations: Before falling asleep I was thinking about the diet I'd started a few days previously, and also about the possibility of going back to school and taking a degree in some area of humanistic psychology.

Dream theme: Conflict about weight.

Puns: Making a point—my breasts are trying to "make a point."

Dream symbols:
Nan — the essence of an attractive, sexy woman who spoiled herself by dieting. The last time I saw her she was unattractively thin.
Breasts — part of me that I value.
Topless bathing suit — sexy.
Boxer shorts — unfeminine.
Tan — sexually attractive.
Diet — the way for me to become thinner and thereby sexually attractive.

Dream message: I obviously have a deep confusion about size and femininity and fear that if I lose weight, I shall become sexually unattractive. This confusion may account for the fact that I slip from my dieting so often.

Action: Get myself together on this issue, though at the moment I don't really have enough information to work on.

17th January, 1973 *15*

Dream: Elizabeth X has picked me up to take me to the airport to catch a plane to Boston. She has directions, but for some unknown reason, we go in the wrong direction. I ask her what she's doing, as we don't have an awful lot of time. She says, "Oh, don't you remember, we agreed I wouldn't take you to the airport but to this other place?" I reply, "No, I don't remember at all." I wonder about getting out of the car and going back, but I don't.

Then we get to the place she is taking me, and it turns out to be a house where a human potential meeting is to be held. It is situated on a lower level, and just as I start down the steps to the entrance, a very unruly dog pulls on one of my sandals and starts to gnaw the sole so that I am unable to get free.

Day's events and associations: A most unpleasant staff meeting at work the previous day, which made me seriously contemplate leaving my job in the very near future.

Dream themes: a. Going in the wrong direction.
b. Pushing away an animal.

Puns: a. I am being "taken for a ride" or being "driven to face" something.
b. I approach a "lower level" of my personality.
c. A dog "gnaws at my soul."
d. Human potential meeting—my own "human potential."

Dream symbols:
Boston — school that has humanistic psychology program.
Elizabeth X — church friend who became a leader in human potential movement (my own inner human potential guide).
Dog — my animal nature—my energy and strength.
Sole — soul.

Notes: The nightmarish staff meeting upset me so badly that I seriously thought of giving notice. You could say I was hurting so badly that I didn't know in what direction I was going. As I haven't seen Elizabeth for some time, she probably represents my own inner guide to deeper levels of the personality and my own energy and strength. I had not attended any human potential meeting recently.

Dream message: My own inner wisdom is driving me to discover my own inner potential, and it seems to be saying that I should do this by working in a group of some kind, and not going to school and getting more degrees and qualifications. This will not solve my problems. The dream also shows that as I go deeper within myself, the first encounter will be with my animal nature, which has become unruly on account of neglect. The work situation is making underdog insistent that I leave and pay attention to him.

Action: I wish I could report that I gave notice at work, but I did not in fact leave until four months later, under hostile and miserable conditions of a court case. [See the dream of Hal in the section on Sex Dreams.]

Permanent dream glossary: The theme of pushing away an animal has recurred and so takes its place in my dream glossary—also the dog, which represents another facet of my animal nature. Also, sole—soul.

21st May, 1973 *25*

Dream: The dream is set in a beautifully landscaped backyard of an anonymous female friend. The grounds are exquisitely beautiful. One or two other people are present, walking around enjoying the unusual flowers, shrubs, and trees. There are also songbirds, purchased by my friend, which she had added to the yard to give additional pleasure. Somehow, they do not fly away. Finally there are two dogs. One is a large red mutt which is very much attached to me. He climbs into my lap, licks my face, and lies at my feet. The other dog is small and basically part of the setting.

My friend and I are seated in chairs under a large tree. The other two people are at a distance, standing and talking. My friend remarks that she has just gotten two new birds. They are lovely, small birds with pale blue and green feathers. She hopes we will enjoy their song. I am looking up at the birds, admiring the colors of their feathers. Just then, one of the other people calls my friend's attention to a rustling in the tree. She says, "Oh, yes, and you must see my 'sunshine monkey.' " A small, pale yellow monkey drops from the tree into her arms. She holds it for a few minutes and then puts it on the ground. Both dogs come and sniff at it in a disinterested manner and return to where they were. The monkey seems alone and bewildered.

Day's events and associations: I dreamed this the night before an interview for a humanistically oriented course at the University of California. I was excited by the program, particularly on account of the possibility of therapy included in it. I love California and had already considered moving there. I have also been much aware of the many beautiful and unusual trees there.

Dream theme: Stranger in paradise.

Puns: a. The big red mutt is "attached" to me (part of me).
b. The garden is mainly "for the birds."

Dream symbols:
Backyard — secluded, away from the traffic and movement of people (my own unperceived and unrecognized psychic space).
Female friend — stranger to me: part of myself of which I am unaware.
Big red dog — my animal nature—underdog—affectionate, caring, attached, and at home in paradise.
Small dog — emphasizes the point that dogs are at home in the garden.
Unusual and beautiful flowers and trees — California.
Birds — "for the birds."
Sunshine monkey — me: part of myself who is a stranger in paradise (my father calls me "monk" for monkey).

Notes: As I have no friends in California, the anonymous female friend must be a part of myself, a stranger to me, who lives in the "backyard" of my psyche, away from all conscious concerns of work, money, strife, etc. She is bringing my lost childhood capacity for enjoyment to my attention and wants me to share it. This paradise is also the home of underdog, who once again wants to make friends with me. My father used to call me "monk" when I was a child, so the sunshine monkey is the part of myself derived from him—undemonstrative, hard-working, and efficient—who is most uncomfortable and out of place in this paradise. As a child, I learned that whatever one makes for oneself, earns, works for, is always better and more enjoyable than something that comes free. As I haven't done any of these things, then it must be "for the birds."

Dream message: I can enjoy paradise only if I learn to accept good gifts from myself (it is your Father's good pleasure to *give* you the Kingdom) and share my life with the innately happy, instinctual, loving part of myself. The driving, striving aspect of my personality (my father's daughter) is new to this kingdom and bewildered by it, but that part of me will learn to live in paradise with care, love, and attention. Paradise is not "for the birds": it is *for me* and for anyone else who chooses to accept and enjoy it, if only we give ourselves permission.

Action: Give myself permission. Change my attitude to life.

Transforming symbol: Paradise—the kingdom of heaven within.

Permanent dream glossary:
a. Big red mutt—loving under dog again.
b. Monkey—stranger in para dise.

22nd June, 1973 *30*

Dream: I have just had surgery and both breasts have been removed. I had not expected anything so radical, and a number of people on the hospital staff are also very upset at what the doctor has done. At the desk near my room one of the research girls is quite hysterical. The situation is reported to upper levels of management, and a man comes and reprimands this young woman. I am able to communicate with the girl nonverbally, to acknowledge my appreciation of her concern and feelings so that she is able to stay together. She controls herself and the situation smooths over.

Day's events and associations: Got a call to come for an interview for a research position with a management firm. It will probably involve a $2,000 cut in pay.

Dream theme: A surgical operation (recurring theme).

Puns: a. Taking a "cut"—salary cut.
b. "Upper levels"—conscious mind.

Dream symbols:
Breasts — an essential and valued part of my body.
Research girl — me (in context of possible new job).
Management — the firm with which I have interview scheduled. Also the part of me that exercises control.

Dream message: I am far more upset about the possible salary cut than I consciously realized—in fact, one part of me is almost hysterical about it. I controlled my feelings so successfully that I never even fully registered them. The message seems very clear that I should be most wary of taking a salary cut. A principle of far greater value than money is involved, and I feel I would lose an essential part of my personal/professional self-image—or something—if I take this position at this salary level, just because it's the first response I've had. The job sounded very boring anyway, and my feeling is that I should be paid *more,* not less, for doing an uninteresting job.

Action: I canceled the interview.

Permanent dream glossary:
Breasts — my feeling of worth or value.
Level — level of the mind.
Surgical operation — a "cutting out operation" which diminishes my worth.

29th December, 1973 37

Dream: I have a small red-haired child with me—a boy, I think. He is two or three years old. Several adults are present and we are discussing the thievery of small children. The adults disappear from the dream and there appears a bowl of fruit on the table. The child takes an orange from the bowl. Considering the conversation that has just taken place, I am amazed at the child's behavior and try to discover why he has taken the fruit. He refuses to put the orange back and insists that he would take the fruit again in the same circumstances.

At a loss what to do, I decide the child must be spanked, and proceed to administer the spanking, although I have little strength in my right arm and my wrist and the back of my hand feels stiff and painful. The child persists in his attitude and I realize that this method of dealing with him is ineffective. A tree without leaves appears. It is glistening wet from rain or sleet. I break off a thin bough and snap it in the air once or twice. I realize that it will not serve, as it disintegrates in my hand from the wetness. As the dream ends, the child has become small, fitting into the palm of my hand. He is holding the orange in both hands and pressing it to his mouth. His face is scrunched up as though about to cry.

Day's events and associations: At home all day. Worked on my dreams and thought a lot about the hard time I give my dog, or underdog. The day before, Mother was recalling the time my father gave me a one-blow spanking when I, at about the age of the child in the dream, started to throw a tantrum. Mother did occasionally spank me and sometimes used a switch to punish me. As an infant, my hair was red and my parents wanted a boy.

Dream theme: Spanking a child for stealing.

Puns: The child steals the "forbidden fruit."

Dream symbols:
Red-haired child — me, underdog, the inner child.
Thievery — in this case, mischief and defiance—potentially serious, but not necessarily.
Bowl of fruit — horn of plenty: plentifulness, bounty, "forbidden fruit."
Orange — one of my favorite fruits.
Wet, leafless tree — we've had rain, snow, and sleet. I think this just places the dream in the present time.
Tiny child — my own human potential.

Notes: The horn of plenty—the forbidden fruit—reminds me of the "sunshine monkey" dream in which I was unable to accept the bounty of paradise where underdog would be at home and wouldn't even hurt the monkey. Now underdog makes it quite clear that he is going to take the fruit and is even prepared to bring in a secret saboteur who weakens my arm and disintegrates the switch! In dialogue, when I expressed anxiety about stealing, the child told me that there were plenty of oranges for everyone and that I needn't fear his becoming a criminal. He told me that my own crazy, mixed-up parental values were responsible for the unnecessary fuss I was making.

Dream message: I am still rough in my handling of the child within, but the dream makes clear the ineffectiveness of my methods and the strength of the child to withstand them. Topdog's power over me is weakening. The reappearance of the tiny child at the end indicates the critical nature of the struggle and the importance of satisfying the child's needs.

Action: Constant daily care and attention to the tiny child—my Christ child within—through dream work, group therapy (if necessary), and above all self-reflection and attention to my inner needs in waking life.

Recurring Symbols from the above dreams as they appear in my Permanent Dream Glossary (in alphabetical order):
Animal — my own animal nature.

Dog — my energy and strength; my loving, affectionate aspect.

Elephant — my overweight self: playful and loving.

Monkey — part of me derived from father, unable to play and enjoy life.

Breasts — my feeling of worth.

Level — level of mind.

Upper level — conscious mind.

Lower level — unconscious mind.

Sole — soul.

Surgical operation — a "cutting out" or "cutting down" operation that I feel diminishes my worth in some way.

Recurring Dream Themes

Being a passenger in a car — relinquishing the "driving" role.

Making friends with an animal — accepting underdog.

Pushing away an animal — rejecting underdog.

Many of the other symbols appearing in the preceding dreams have recurred and been recorded in my permanent dream glossary. Some of the symbols and their meanings are: tennis shoes (large soul in contact with the earth); Nan (my sexy, competitive aspect); a child (my inner child—underdog); tiny baby (my own human potential); thievery (removal of topdog's values). Sometimes a recurring symbol will not have its usual meaning, but this is rare. The time saved by compiling and using a dream glossary is enormous, and some dreams virtually interpret themselves once the art has been mastered.

Since writing this, I have had two more dreams which throw light on my dieting problem. Having failed on one diet, I contemplated a more rigorous one, and the night before I was due to start it I dreamed of being carved up and eaten by Chinese cannibals! I should have heeded the dream's warning to forget this particular diet, but I persisted and after two days came down with such severe diarrhea that I had to quit. This astonished me as the warning with the diet was one of possible constipation. Resuming my standard low-calorie diet on my recovery two weeks later, I failed to lose weight as I normally do, and so I contemplated a fast. I then dreamed of someone carving flesh from my left side! This time, I took the warning and abandoned the notion of a fast, for clearly underdog sees any kind of rigorous diet as topdog's onslaught to "cut him down to size" or "diminish his stature" in some way. The result is sabotage, and I have now come to the conclusion that dieting without further self-therapy is useless. It is not a question of will power—a meaningless term—but of deep dynamics in the psyche. I believe that as I gradually clear out my topdogs and come to terms with my underdog soul, my body will take care of itself. And I'm going to start by finding my way back to the Garden, which is underdog's natural home, and eating as much "forbidden fruit" as I can manage. That's the best slimming diet I know!

BIBLIOGRAPHY

Basic Recommended Reading on Dreams for the General Reader

Bro, Harmon H. *Edgar Cayce on Dreams.* New York: Paperback Library, 1968.

Diamond, E. *The Science of Dreams.* Garden City, N.Y.: Doubleday, 1962.

Faraday, Ann. *Dream Power.* New York: Coward, McCann & Geoghegan, 1972; Berkley Medallion Books, 1973.

Hall, Calvin S. *The Meaning of Dreams.* New York and London: McGraw-Hill, 1966.

———, and Nordby, Vernon. *The Individual and His Dreams.* New York: Signet Books, 1972.

Jung, C. G. et al. *Man and His Symbols.* London: Aldus Books, 1964; New York: Dell, 1968.

Luce, G., and Segal, J. *Sleep.* New York: Coward, McCann & Geoghegan, 1966.

MacKenzie, N. *Dreams and Dreaming.* London: Aldus Books, 1966.

Perls, Frederick S. *Gestalt Therapy Verbatim.* Moab, Utah: Real People Press, 1969.

Sanford, John A. *Dreams: God's Forgotten Language.* Philadelphia and New York: J. B. Lippincott, 1968.

Sechrist, Elsie. *Dreams—Your Magic Mirror.* New York: Dell, 1969.

Other Books Cited or Quoted in the Text

Alpert, Richard (Baba Ram Dass). *Be Here Now.* New York: Crown Publishers, 1971.

Berne, Eric. *Games People Play.* New York: Grove Press, 1964; paperback, 1967.

Caligor, Leopold, and May, Rollo. *Dreams and Symbols: Man's Unconscious Language.* New York and London: Basic Books, 1968.

Castaneda, Carlos. *Journey to Ixtlan.* New York: Simon & Schuster, 1972; paperback, 1973.

Devereux, G., ed. *Psychoanalysis and the Occult.* New York: International Universities Press, 1953.

Downing, Jack, and Marmorstein, Robert. *Dreams and Nightmares: a Book of Gestalt Therapy Sessions.* New York: Harper & Row, 1973; Perennial Library, 1973.

Dunne, J. W. *An Experiment with Time.* London: Faber & Faber, 1927; New York: Hillary House, 1958.

Ehrenwald, Jan. *New Dimensions of Deep Analysis.* New York: Grune & Stratton, 1954.

Fox, Oliver. *Astral Projection.* New Hyde Park, N.Y.: University Books, 1962.

Friedan, Betty. *The Feminine Mystique.* New York: Norton, 1963.

Fromm, Erich. *The Forgotten Language.* New York: Holt, Rinehart & Winston, 1951.

Green, Celia. *Lucid Dreams.* Oxford: Institute of Psychophysical Research, 1968.

Gregg, Douglas M. *Hypnosis, Dreams and Dream Interpretation* (booklet). San Diego, Calif.: Medical Hypnosis Center, 1970. (Obtainable from 1767 Grand Ave. Suite 107, San Diego Ca. 92109)

Harris, Thomas A. *I'm OK—You're OK: A Practical Guide to Transactional Analysis.* New York: Harper & Row, 1967; Avon, 1973.

Hartmann, E., ed. *Sleep and Dreaming.* Boston: Little, Brown, 1970.

Huang, Al. *Embrace Tiger, Return to Mountain—the Essence of T'ai Chi.* Moab, Utah: Real People Press, 1973.

Illich, Ivan. *Tools for Conviviality.* New York: Harper & Row, 1973.

Janov, Arthur. *The Primal Revolution.* New York: Simon & Schuster, 1972; Touchstone paperback, 1973.

Jourard, Sidney M. *The Transparent Self.* New York and London: Van Nostrand Reinhold, 1971.

Jung, C. G. *Two Essays in Analytical Psychology: Collected Works vol. 7; The Structure and Dynamics of the Psyche: C. W. vol. 8; Civilization in Transition: C. W. vol. 10; Psychology and Alchemy: C. W. vol. 12; Alchemical Studies: C. W. vol. 13.* Princeton: The University Press (Bollingen Series); London: Routledge & Kegan Paul.

Koestler, Arthur. *Insight and Outlook.* Lincoln: University of Nebraska Press, 1965 (paperback).

Kreskin. *The Amazing World of Kreskin.* New York: Random House, 1973.

Logan, Daniel. *The Reluctant Prophet.* New York: Doubleday, 1968; Avon, 1969.

Masters, Brian. *Dreams about Her Majesty the Queen and other Members of the Royal Family.* London: Blond & Briggs, 1972; Mayflower Books, 1973.

Masters, Robert, and Houston, Jean. *The Varieties of Psychedelic Experience.* New York: Holt, Rinehart & Winston, 1966; Dell, 1966.

———. *Mind Games.* New York: Viking, 1972.

Meier, C. A. *Ancient Incubation and Modern Psychotherapy.* Evanston, Ill.: Northwestern University Press, 1967.

Merton, Robert K. *The Sociology of Science.* Chicago: The University Press, 1973.

Miller, Stuart. *Hot Springs.* New York: Viking, 1971.

Noone, Richard. *In Search of the Dream People.* New York: Morrow, 1972.

Pearce, Joseph Chilton. *The Crack in the Cosmic Egg.* New York: Julian Press, 1971.

Roszak, Theodore. *Where the Wasteland Ends.* New York: Doubleday, 1972; Anchor, 1973.

Sabini, Meredith. *The Dream Group: a Community Mental Health Proposal.* Doctoral dissertation presented to the California School of Professional Psychology, 1972, and published on demand by Xerox University Microfilm Service, Ann Arbor, Michigan.

Seaman, Barbara. *Free and Female.* New York: Coward, McCann & Geoghegan, 1972.

Stekel, Wilhelm. *The Interpretation of Dreams.* New York: Liveright, 1943.

Stewart, Kilton—see Tart, below.

Tart, Charles T., ed. *Altered States of Consciousness.* New York and London: John Wiley, 1969.

Tillich, Hannah. *From Time to Time.* New York: Stein & Day, 1973.

Ullman, Montague, and Krippner, Stanley, with Vaughan, Alan. *Dream Telepathy.* New York: Macmillan, 1973.

Van Dusen, Wilson. *The Natural Depth in Man.* New York: Harper & Row, 1972.

Vilar, Esther. *The Manipulated Man.* New York: Farrar, Straus & Giroux, 1972; Bantam, 1974.

Von Grunebaum, G. E., and Caillois, Roger, eds. *The Dream and Human Societies.* Berkeley and Los Angeles: University of California Press, 1966; Cambridge: The University Press, 1966.

Further Books of Interest

Agee, Doris. *Edgar Cayce on E.S.P.* New York: Warner Paperback Library, 1969.

Assagioli, Roberto. *Psychosynthesis.* New York: Viking, 1971.

Bonime, Walter. *The Clinical Use of Dreams.* New York: Basic Books, 1962.

Boss, M. *The Analysis of Dreams.* New York: Philosophical Library, 1958.

Campbell, Joseph, ed. *Myths, Dreams and Religion.* New York: Dutton, 1970.

Castaneda, Carlos. *The Teachings of Don Juan.* Berkeley and Los Angeles: University of California Press, 1968; Ballantine, 1969; University of California Press paperback, 1973.

———. *A Separate Reality.* New York: Simon & Schuster, 1971; Pocket Books, 1972.

Cayce, Hugh Lynn. *Venture Inward.* New York: Harper & Row, 1972.

———. *Dreams: The Language of the Unconscious.* Virginia Beach: ARE Press, 1962.

De Becker, Raymond. *The Understanding of Dreams: or the Machinations of the Night.* New York: Hawthorn Books, 1968.

Dunne, J. W. *An Experiment with Time.* New York: Hillary House, 1958.

Eisenbud, Jule. *Psi and Psychoanalysis.* New York: Grune & Stratton, 1970.

Eliade, Mircea. *Myths, Dreams and Mysteries.* New York: Harper & Row, 1960.

Fodor, Nandor. *New Approaches to Dream Interpretation.* New York: Citadel Press, 1951.

Foulkes, David. *The Psychology of Sleep.* New York: Charles Scribner's Sons, 1966.

French, Thomas M., and Fromm, Erika. *Dream Interpretation: a New Approach.* New York and London: Basic Books, 1962.

Freud, Sigmund. *The Interpretation of Dreams.* London: Hogarth Press, 1953; New York: Basic Books, 1965; Avon, 1967.

Green, Celia. *Out of the Body Experiences.* London: Hamish Hamilton, 1968.

Hadfield, J. A. *Dreams and Nightmares.* Baltimore, Md.: Penguin Books, 1954.

Hall, Calvin S. *A Primer of Freudian Psychology.* New York: Mentor Books, 1954.

———, and Nordby, Vernon. *A Primer of Jungian Psychology.* New York: Mentor Books, 1973.

———, and Lind, Richard E. *Dreams, Life and Literature: a Study of Franz Kafka.* Chapel Hill, N. C.: University of North Carolina Press, 1970.

———, and Van de Castle, Robert L. *The Content Analysis of Dreams.* New York: Appleton-Century-Crofts, 1966.

Hill, Brian. *Gates of Horn and Ivory: An Anthology of Dreams.* New York: Taplinger, 1967.

Jacobi, Jolande. *C. G. Jung: Psychological Reflections.* Princeton, N. J.: Princeton University Press (Bollingen Series), 1970.

Jaffe, Aniela. *From the Life and Work of C. J. Jung.* New York: Harper & Row, 1971.

James, Muriel, and Jongeward, Dorothy. *Born to Win: Transactional Analysis with Gestalt Experiments.* Reading, Mass: Addison-Wesley, 1973.

Janov, Arthur. *The Primal Scream.* New York: Dell Publishing, 1970.

Jones, R. M. *The New Psychology of Dreaming.* New York: Grune & Stratton, 1970; Viking Compass edition, 1974.

Jourard, Sidney M. *Healthy Personality: An Approach from the Viewpoint of Humanistic Psychology.* New York: Macmillan, 1974.

Jung, C. G. *The Portable Jung,* Joseph Campbell, ed. New York: Viking, 1971.

———. *Dreams* (selections from the Collected Works). Princeton, N. J.: Princeton University Press, 1974.

———. *Memories, Dreams, Reflections.* New York: Random House, 1961; Vintage Books, 1962.

Kelsey, Morton T. *Dreams: Dark Speech of the Spirit.* Minneapolis, Minn.: Augsburg, 1968; paperback edition *God, Dreams and Revelation,* 1974.

———. *Encounter with God.* Minneapolis, Minn.: Bethany Fellowship, 1972.

Kerouac, Jack. *Book of Dreams.* San Francisco, Calif.: City Lights, 1961.

Koestler, Arthur. *Insight and Outlook.* Lincoln: University of Nebraska Press, 1965 (paperback).

Kramer, M., ed. *Dream Psychology and the New Biology of Dreaming.* Springfield, Ill.: Charles C. Thomas, 1969.

LeShan, Lawrence. *The Medium, the Mystic and the Physicist: Towards a General Theory of the Paranormal.* New York: Viking, 1974.

Lilly, John C. *The Center of the Cyclone: an Autobiography of Inner Space.* New York: Julian Press, 1972.

Martin, P. W. *An Experiment in Depth.* New York: Humanities Press, 1955; London: Routledge & Kegan Paul, 1955.

Monroe, Robert A. *Journeys out of the Body.* New York: Doubleday, 1972; Anchor, 1973.

Oswald, Ian. *Sleep.* Harmondsworth and Baltimore, Md.: Penguin, 1966.

Perls, Frederick S. *In and Out of the Garbage Pail.* Moab, Utah: Real People Press, 1969.

Priestley, J. B. *Man and Time.* London: Aldus Books, 1964.

Psycho-Sources: a Psychology Resource Catalog. New York and London: Bantam, 1973.

Rawson, Wyatt. *The Way Within.* London: Vincent Stuart, 1963.

Roheim, Geza. *The Gates of the Dream.* New York: International Universities Press, 1952; paperback, 1969.

Rossi, Ernest Lawrence. *Dreams and the Growth of Personality.* New York and Oxford: Pergamon Press, 1972.

Roszak, Theodore. *The Making of a Counter-Culture.* New York: Doubleday, 1969; Anchor, 1969.

______. *Sources.* New York: Harper & Row, 1972.

Rycroft, Charles, ed. *Psychoanalysis Observed.* New York: Coward, McCann & Geoghegan, 1967. Baltimore, Md.: Penguin, 1968; London: Constable, 1966; Harmondsworth: Penguin, 1968.

Schutz, William C. *Joy: Expanding Human Awareness.* New York: Grove Press, 1967; Black Cat edition, 1969.

______. *Here Comes Everybody.* New York: Harper & Row, 1971.

Stearn, Jess. *Edgar Cayce, the Sleeping Prophet.* New York: Doubleday, 1967; Bantam, 1968.

Stevens, Barry. *Don't Push the River.* Moab, Utah: Real People Press, 1970.

Stevens, John O. *Awareness.* Moab, Utah: Real People Press, 1971.

Stewart, Kilton. *How to Educate Your Dreams to Work for You* (booklet). New York: The Stewart Foundation for Creative Psychology, 144 E. 36th St., N.Y., N.Y. 10016.

______. *Dream Theory in Malaya*—see Tart, above.

———. *Dream Exploration among the Senoi*—see Roszak, *Sources,* above.

Van de Castle, Robert L. *The Psychology of Dreaming.* New York: General Learning Corporation, 1971.

Witkin, H. A., and Lewis, H. B., eds. *Experimental Studies of Dreaming.* New York: Random House, 1967.

Wren-Lewis, John. *Love's Coming of Age*—see Rycroft, above.

INDEX

Al Huang, 215–216
Alarm clock experiment, 47
Amazing World of Kreskin, The (Kreskin), 328
Animal nightmares, 241–245
Anniversary dreams, 16, 40
Anxiety, 34, 83
Aquinas, Thomas, 67
Archetypal symbols, 35, 63–65, 293, 304
Assaglioli, Roberto, 13
Association for Humanistic Psychology, 34
Association for Research and Enlightenment (A.R.E.), 102
Astral projection. *See* Out-of-the-body experiences
Astral Projection (Fox), 341
Attack, dreams of, 236–241
Auto-hypnotic procedure, 41
Autosuggestion, 143

Bacon, Francis, 169
Becker, Ernest, 165
Berne, Eric, 33, 34, 171
Blake, William, 194, 199, 243, 299, 310, 331
Body language, 60
  dream puns based on, 107–111
Brain-wave records, 17–18
Bro, Harmon, 36
Brooks, John Benson, 161

Carson, Johnny, 71
Cass, Peggy, 70
Castaneda, Carlos, 25–26, 340–341
Castration anxiety, 78
Catastrophic expectations, 29–32, 182, 222, 224, 225, 251, 279, 338
Cayce, Edgar, 9, 11, 13, 15–16, 36, 62, 94, 158, 270, 287, 328
  on dream clarification, 145–147
  on paralysis dreams, 252
Cayman Islanders, 62
Chaucer, Geoffrey, 231, 252
Cicero, 314
Clairvoyance, 38, 311, 316, 319
Clarifying essays, 145–147
Clark, Kenneth, 322–323
Colloquial metaphors, 103–107
Color, in dreams, 44–45
Counterculture, 200–201, 243
Creative inspiration, 160–165
Creative monologue technique, 129, 158
Cryptomnesia, 161

Daydreams, 10
Dean, Jimmy, 72
Death dreams, 267–289
of the dead, 273–281
death and rebirth of self, 282–289
of family, 268–270
of strangers, 270–273
Dench, Judi, 219
Devereux, George, 318
Digestive system, 47
Disguise theory, 53–55
Douglas, Mike, 320
Downing, Jack, 226
Dream books, 59
Dream cartoons, 103–107
Dream diary, 48, 312–316, 322, 323
Dream fragments, 38, 39, 300
Dream glossary, 73, 115, 116
Dream power
and creative inspiration, 160–165
as dream interpreter, 143–148
as family counselor, 150–152
resolving nightmares by, 255–258
as spiritual counselor, 156–160
as *spiritus rector,* 148–150
as vocational counselor, 152–156
"Dream Psychology of the Senoi Shaman, The" (H. D. Noone), 259
Dream Telepathy (Ullman, Krippner, and Vaughan), 311
Dream workshops, 27, 32, 129, 183–184, 186
Dreams. *See also* Nightmares
anniversary, 16, 40
clarifying essays, 145–147
color in, 44–45
death in. *See* Death dreams
digestive system and, 47
drama of, 7–9
eliciting help from. *See* Dream power
extrasensory perception and, 38, 271, 310–336
Freudian theory. *See* Freud, Sigmund
Jungian theory. *See* Jung, Carl Gustav
lucid, 26, 57, 263–266, 286, 334–343
nonrecall of, 19–21, 25, 27–32, 35–36
out-of-the-body experiences, 287, 340–343
as picture language, 4, 51–56
positive function of, 10–12
prior events and, 3–4, 15–17
puns in. *See* Puns
recall of, 21–22, 27, 37–47
recording, 37–47
recurring, 68, 106
self-reflection, 337
symbols in. *See* Symbolism, dream
themes. *See* Themes, dream
topdog/underdog conflict. *See* Topdog/underdog model of the personality
unconscious and, 12–15
Dreams: God's Forgotten Language (Sanford), 234, 241
Dreams of Her Majesty the Queen and Other Members of the Royal Family (Masters), 127
Dreams and Nightmares (Downing), 226
Dreams—Your Magic Mirror (Sechrist), 102, 158
Drowning, dreams of, 246–250
Drugs, psychedelic, 18, 89, 329
Dunne, J. W., 313–314, 316

Edgar Cayce on Dreams (Bro), 36

Ehrenwald, Jan, 317–318
Einstein, Albert, 319
Eisenbud, Jule, 335
Embrace Tiger, Return to Mountain (Al Huang), 216
Examination dreams, 75–78
Experiment with Time, An (Dunne), 313
Extrasensory perception (ESP), 38, 271, 310–336

Falling, dreams of, 69–70
Family life, 193–194
Famous people, dreams of, 124–128
Fantasy, 30
Feminine Mystique, The (Friedan), 210
Flying, dreams of, 71–72
Forgetting lines, dreams of, 219–226
Forgotten Language, The (Fromm), 114
Fox, Oliver, 341
Free and Female (Seaman), 162
Freud, Sigmund, 93, 94, 101, 104, 170–171, 320
  disguise theory, 53–55
  on dream symbolism, 60–61, 63, 64
  extrasensory perception and, 316–317, 328
Friedan, Betty, 210
From Time to Time (Tillich), 269
Fromm, Erich, 114
Frost, David, 102
Frustration dreams, 204–209
  forgetting lines and inability to perform, 219–226
  hidden heroes, 229
  lateness and unpreparedness, 226–228
  missing vehicles, 207–213
  telephoning, 213–216
  toilet thief, 228–229
  vanishing objects, 216–218

Garrett, Eileen, 328
Gestalt therapy, 36, 121, 170, 180, 183, 202–203, 234
Gestalt Therapy Verbatim (Perls), 195, 222, 262, 281
Goethe, Johann Wolfgang von, 267
Goldsmith, Joel, 338
Gordon, Bill, 100–101
Green, Celia, 341
Gregg, Douglas M., 156, 157, 160

Hall, Calvin, 57–58, 66, 158, 159, 233, 247, 271
Hammel, Lisa, 104
Hardy, Sir Alister, 328
Harris, Thomas, 285–288, 329
Hartman, Ernest, 231
Heywood, Rosalind, 328
High lucid dreams, *see* Lucid dreams
Hope, Bob, 102
Hot Springs (Miller), 224
Houston, Jean, 165, 329
Howe, Elias, 160
Hypnosis, Dreams and Dream Interpretation (Gregg), 156

Id, 170
Illich, Ivan, 33
I'm OK—You're OK (Harris), 285–286, 288
In Search of the Dream People (Richard Noone), 262
Inability to perform, dreams of, 219–226
Incubus attacks, 231–232
Indigestion, 47
Insight and Outlook (Koestler), 94
International Psychiatry Clinics, 231

Interpretation of dreams. *See* Puns; Symbolism, dream; Themes, dream; Topdog/underdog model of personality
Interpretation of Dreams, The (Stekel), 157
Intoxication, 47
Intrusion, dreams of, 236–241

Janov, Arthur, 279–280
Joplin, Janis, 89
Jourard, Sidney, 308
Journey to Ixtlan (Castaneda), 26, 179, 340
Joyce, James, 78–79
Juan, Don, 25–27, 179, 335, 340–341
Jung, Carl Gustav, 5, 12, 13, 43, 54, 66, 94, 113, 117, 138, 140, 187, 229, 261, 274, 281, 302
  on archetypal symbolism, 63–64, 293
  on extrasensory perception, 326–327
  on nightmares, 234
  transcendent function, 295–296

Katz, Leo, 162–163
Kekule von Stradonitz, Friedrich, 160
Kelly, 169
Koestler, Arthur, 20, 94
Kreskin, 162, 321, 322, 328
Krippner, 311, 312

Lamb, Charles, 92
Lamour, Dorothy, 102
Lao-tse, 178–179
Lapis, The (Pereira), 163–164
Lateness, dreams of, 226–228
Loewi, Otto, 161
Logan, Daniel, 328
Loneliness, 8, 83
Looking-inward dreams, 139
Looking-outward dreams, 138–139
Looking-glass dreams, 139
Lucid dreams, 26, 57, 263–266, 286, 334–343
Lucid Dreams (Green), 341
Lynn, Hugh, 48, 270

Manipulated Man, The (Vilar), 210
Marriage of Heaven and Hell, The (Blake), 243
Masters, Brian, 127, 128
Masters, Robert, 165, 329
May, Rollo, 112
Mead, Margaret, 111, 160
Meaning of Dreams, The (Hall), 57, 233, 247
Mearns, Hughes, 204
Meditation, 23
Memory, 16, 38, 42
Merton, Robert, 161
Metamorphosis 1942 (Katz), 162
Miller, Stuart, 224
Milton, John, 185
Missing vehicles, dreams of, 207–213
Money and valuables, dreams of, 79–83
Moore, Jack, 169, 176
Muggeridge, Malcolm, 87
Mysticism, 23

Natural Depth in Man, The (Van Dusen), 152, 330
Nebel, Long John, 92–93
New Dimensions of Deep Analysis (Ehrenwald), 317
Nightmares, 3, 32, 35, 68, 231–266
  animal, 241–245
  conflicting views on, 233–234
  incubus attacks, 231–232
  lucid, 263–266
  paralysis, 250–252
  of pursuit, attack, violence, and

Nightmares *(cont'd)*
intrusion, 236–241
recurring, 232–233, 242–243
resolving, 252–266
self-punishment, 233–235
tidal waves and drowning, 246–250
Non-rapid eye movement (NREM) sleep, 20, 47, 250
Nonrecall of dreams, 19–21, 25, 27–32, 35–36
Nonsymbolic dream characters, 118–122
Noone, H. D. ("Pat"), 258, 259, 262, 266
Noone, Richard, 262
Nudity, dreams of, 73–75

Occult groups, 24–25
Oneirolysis, 20
Out-of-the-body experiences, 287, 340–343

Paracelsus, 3
Paradise Lost (Milton), 185
Paralysis nightmares, 250–252
Pascal, Blaise, 6
Paul, Saint, 177
Peanuts, 9–10
Pearce, Joseph Chilton, 290
Pereira, Irene Rice, 163–165
Perfectionism, 223–224, 299
Perls, Fritz, 29, 34, 36, 111, 138, 222, 261, 262, 281, 289, 295, 326
topdog/underdog model, 87, 170–171, 176, 180–182, 184, 189–191, 195, 200, 202–203
Picture language, dreams as, 4, 51–56
Precognition, 311, 312, 313, 316, 319
Pre-lucid dreams, 337, 338
Priestley, J. B., 313
Primal Revolution, The (Janov), 279–280
Psychedelic drugs, 18, 89, 329
Psychoanalysis, 24, 93, 210–211, 318
Psychoanalysis and the Occult (Devereux), 318
Psychosynthesis, 13
Psychotherapy, 15, 24, 32–34, 170, 211, 279
Puns, 60, 91–111
body language and, 107–111
dream cartoons, 103–107
finding, 111
proper names, 100–103
reversal, 97–98
verbal, 95–97
visual, 98–100
Pursuit, dreams of, 236–241

Rapid eye movement (REM) sleep, 19–20, 47, 165, 231, 232, 250, 310, 312
Recall of dreams, 21–22, 27, 37–38
instructions for, 39–47
Recording dreams, 37–38
instructions for, 39–47
Recurring dreams, 68, 106
Recurring nightmares, 232–233, 242–243
Relaxation exercise, 41
Reluctant Prophet, The (Logan), 328
REM sleep. *See* Rapid eye movement (REM) sleep
Reversal puns, dreams based on, 97–98
Rhine, J. B., 312
Rogers, Carl, 34
Roszak, Theodore, 19, 23, 335

Sabini, Meredith, 160–161
Sanford, John, 234, 241

Seaman, Barbara, 161–162
Sechrist, Elsie, 102, 158, 273, 274
Secondary gain, 157, 159
Secret of the Golden Flower, The (Jung), 296
Self, death and rebirth of, 282–289
Self-punishment nightmares, 233–235
Self-reflection dreams, 337
Servadio, Emilio, 325
Sex dreams, 72, 84–90, 340
Shaw, Artie, 319–321
Slang expressions, 60, 103–107
Sleep and Dreaming (ed. Hartmann), 231
Sleep-talking, 250
Sleep-walking, 250
Snake symbolism, 61–64
Sociology of Science, The (Merton), 161
Stekel, Wilhelm, 157
Stevenson, Robert Louis, 160
Stewart, Kilton, 259, 282
Sun symbolism, 65
Superego, 170, 171
Symbolism, dream, 53–55, 57–65, 112–118
  archetypal, 35, 63–65, 293, 304
  nonsymbolic dream characters, 118–122
  snakes, 61–64
  sun, 65
  symbolic dream characters and images, 122–138
  transforming, 290–309

Teeth, dreams of losing, 78–79
Teilhard de Chardin, Pierre, 302
Telepathy, 311, 312, 316, 318–319
Telephoning, dreams of, 213–216
Temiar tribe, 258–263, 266, 282
Themes, dream, 59–60, 67–91. *See also* Frustration dreams
  examination, 75–78
  falling, 69–70
  flying, 71–72
  identifying, 90–91
  money and valuables, 79–83
  nudity, 73–75
  sex, 72, 84–90, 340
  teeth, 78–79
Tidal waves, dreams of, 246–250
Tillich, Hannah, 269
Tillich, Paul, 269, 289
Tools for Conviviality (Illich), 33
Topdog/underdog model of the personality, 87, 170–203
  death dreams and, 269–270, 279–280, 284, 287–288, 295, 298, 301–304
  extrasensory perception and, 312, 321–329, 334, 336
  frustration dreams and, 206, 207, 209–229
  lucid dreams and, 337, 340, 343
  nightmares and, 234–235, 237–238, 244–245, 248–252, 257–258, 260–261, 263–265
Transactional Analysis, 33, 285, 325
Transcendent Function, 229, 295–296
Transforming symbols, 290–309
Transparent Self, The (Jourard), 308
Turner, Gladys Davis, 244

Ullman, 311, 312
Ulysses (Joyce), 78–79
Unconscious, the, 12–15, 170, 249
Unpreparedness, dreams of, 226–228

Van Dusen, Wilson, 51, 152, 330
Vanishing objects, dreams of, 216–218
Varieties of Psychedelic Experience and Mind Games, The (Mas-

Varieties of Psychedelic (cont'd) ters and Houston), 165, 329
Vasconcellos, John, 201–202
Vaughan, 311
Verbal puns, dreams based on, 95–97
Verification dreams, 160
Vilar, Esther, 210
Violence, dreams of, 236–241
Visual puns, dreams based on, 98–100
Volkan, Vamik, 278

Waking fantasy, resolving nightmares in, 252–255
Where the Wasteland Ends (Roszak), 23
Wilde, Oscar, 37

ABOUT THE AUTHOR

Ann Faraday received her B.Sc. and Ph.D. degrees in psychology from University College, London. After several years in experimental dream research, she trained in hypnotherapy, Freudian and Jungian analysis and Gestalt therapy. Her experience convinced her of the limitations of all dream theories used by psychotherapists, and of the need for a radically new approach whereby the dreamer learns to understand his own unique dream language. From her exploratory dream study groups conducted in the U.S. and abroad, she evolved her well-known three-stage method of dream interpretation, which is now widely used by individuals, families, therapists, church groups and on college campuses.

Dr. Faraday has written and lectured widely on dreams, drugs, hypnosis, altered states of consciousness and ESP, and has conducted dream workshops in many parts of the world. She was a pioneer of the Human Potential Movement and of the Association for Humanistic Psychology in Great Britain, and is an active participant in these movements in the United States. Her first book, *Dream Power,* was published in 1972, and she is currently experimenting with expansion of consciousness through dreams.